Contents

Table A

List of Scientific Terms, Concepts and Principles used in Unit 1

	Taken as prerequisites		Introduced in this Unit			
1 Assumed from general knowledge	2 Introduced in previous Unit	Unit No.	3 Developed in this Unit or in its set book	Page No.	4 Developed in a later Unit	Unit No.
		S100[1]		**Unit**		
	radioisotope	6	electrophoresis	12	active site	2
	molecular weight (MW)	8	gel filtration	10	mitochondrion	3
	electrodes	9	specific activity	9		
	ions	9	criteria of purity	13		
	pH	9	unusual bases	20		
	amino acid	10	endopeptidase	15		
	apolar	10	exopeptidase	16		
	chiral	10	β-sheet	23		
	conformation	10	optical rotatory dispersion (ORD)	27		
	hydrogen bonding	10	base stacking	26		
	isomer	10	hypochromism	26		
	polar	10	side-groups	29		
	cellulose	13	apolar forces	29		
	cross-linking	13	polar side-groups	29		
	glucose	13	apolar side-groups	29		
	glycogen	13	ionic bonding	29		
	hydrolysis	13	random chains	29		
	macromolecule	13	active conformation	30		
	monomer	13	quaternary structure	30		
	peptide bond	13	subunit	30		
	polymer	13	sedimentation equilibrium method	14		
	polysaccharide	13	sedimentation velocity method	14		
	protein	13*				
	α-helix	14*		**CS&F**[2]		
	C-terminus	14*	subunit interactions	223		
	disulphide bond	14*	self-assembly	268		
	membrane	14	aided assembly	268		
	N-terminus	14*	directed assembly	268		
	organelles	14				
	primary structure	14*				
	secondary structure	14*				
	tertiary structure	14*				
	substrate	14				
	ultracentrifuge	14				
	enzymic activity	15				
	nucleic acid	17				
	nucleotide	17				
	DNA	17				
	RNA	17				
	mRNA	17				
	tRNA	17				
	X-ray crystallography	28				

*The terms marked with an asterisk in column 2, which were introduced in S100, Units 13 and 14, have also been developed in this Unit.

[1]*The Open University (1971) S100* Science: A Foundation Course *The Open University Press.*

[2]*A. G. Loewy and P. Siekevitz (1969)* Cell Structure and Function (*2nd ed.*) Holt, Rinehart and Winston.

12 msc

The Open University

Second Level Course

BIOCHEMISTRY

UNIT 1
Macromolecules

UNIT 2
Enzymes

University Course Team

THE OPEN UNIVERSITY PRESS

THE BIOCHEMISTRY COURSE TEAM
Steven Rose (*Chairman and General Editor*)
Norman Cohen
Jeff Haywood
Brian Tiplady
Eve Braley-Smith (*Editor*)
Robin Harding (*Course Assistant*)
Bob Cordell (*Staff Tutor*)
Vic Finlayson (*Staff Tutor*)
Roger Jones (*BBC*)
Jim Stevenson (*BBC*)

The Open University Press
Walton Hall Bletchley Bucks

First published 1972

Copyright © 1972 The Open University

Designed by the Media Development Group of the Open University.

Printed in Great Britain by
Staples Printers Limited
at St Albans

SBN 335 02050 X

This text forms part of the correspondence element of an Open University Second Level Course. The complete list of units in the course is given at the end of this text.

For general availability of supporting material referred to in this text, please write to the Director of Marketing, The Open University, Walton Hall, Bletchley, Bucks.

Further information on Open University courses may be obtained from the Admissions Office, The Open University, P.O. Box 48, Bletchley, Bucks.

Objectives

After you have studied this Unit, you should be able to:

1 Demonstrate your knowledge of the items in Table A by (*a*) choosing the best definitions of them from various alternatives; (*b*) defining them in your own words.

2 Choose the best way of purifying a macromolecule from given experimental data.

3 Deduce the primary structure of a macromolecule from given experimental data.

4 Deduce the quaternary or higher structures of a macromolecule from given experimental data.

5 Deduce the mode of assembly of a macromolecular complex from given experimental data.

Study Guide

This Unit draws quite heavily from S100 and all the concepts, etc., are referenced out to it. We have used a standard system throughout for these references: S100 with a superscript e.g. (S100[18]). If you wish to refresh your memory about any of these terms the exact references can be found at the back of the Unit. The next Section is a Pre-Unit Assessment Test which will allow you to test your recollection of S100.

The set book for the Unit is:
A. G. Loewy and P. Siekevitz (1969) *Cell Structure and Function* (2nd ed.) Holt, Rinehart and Winston from which you will be directed to read:
for Section 1.7, pp. 223–35 (approx. 700 words)
for Section 1.8, pp. 268–85 (approx. 4 000 words)
for Section 1.8.1, pp. 469–80 (approx. 4 500 words)

Other sources of material are:
Steven Rose (1970) *The Chemistry of Life*, Penguin Books: Open University edition.

S. Hurry (1965) *The Microstructure of Cells*, John Murray.

There are also passages in S100, Unit 13 and in *The Chemistry of Life* which you will be directed to read.
These are: S100, Unit 13, Section 13.3.1
          The Chemistry of Life, pp. 45–52 (approx. 1 800 words)

If you wish to organize your study so that you can break off at sensible points in the text we recommend the following:
        at the end of Section 1.2.3;
        at the end of Section 1.6.

The central portion of the Unit, i.e., from Sections 1.3 to 1.6, is the most difficult.

You will be able to test your comprehension of these portions of the Unit by the appropriate Self-assessment Questions at the end of the Unit.

Pre-Unit Assessment Test (for recall of S100)

Because we can make no exact judgment of how much biochemistry you already know, we have had to make certain assumptions about it. Before starting this Unit you need to discover if your knowledge matches our assumptions about it. So test yourself against the eleven questions which follow. After you have finished, check yourself against the answers on p. 38, where you will also find the appropriate references, should you need them.

Answer True *or* False

1   A radioactive isotope is a form of an element with an unstable nucleus.

2   Hydrogen bonds are those bonds which hold hydrogen atoms together in the gaseous form.

3   RNA stores genetic information.

4   A protein is made from amino acids joined by phosphate bonds.

5   A molecule is said to be hydrolysed when it is broken down to give water.

6   The pH of a solution is a measure of the concentration of free protons in it.

7   Polysaccharide is an elaborate name for sugar.

8   The reactions of the electron transport chain are carried out in the mitochondrion.

9   Biological macromolecules are not broken up in the cell.

10   When a molecule is ionized it always gains electrons.

11   Some cells possess no macromolecules.

1.1 Introduction: What are Biological Macromolecules?

Macromolecules are ubiquitous in the biological world; no organism, however simple, is without a multitude of different kinds. It is interesting that those found in the simplest of organisms, the bacteria, are very similar to those in the most complex, the primates. No one has found primitive macromolecules, i.e. ones which are evolutionarily early or crude in their actions.

The present concepts of macromolecular structure only began to emerge about 1930. Before this the large molecules separated from biological materials were thought to be aggregates of relatively small molecules (less than molecular weight 5 000) held together in some loose and unspecified way. It was generally agreed that no substances could exist in the cell which were both large and specific in structure or function. Even as recently as 1920 the cell chemist Hess said that he did not see how a substance which consisted of a very large number of carbon atoms in a long chain could exist in the cell at all, let alone serve any cellular function.

The arguments about macromolecular structure centred around two opposing theories. One, the 'aggregate' theory, proposed that the large molecules were loose aggregates of smaller molecules, whereas the other proposed that a macromolecule could have a specific defined structure. If the molecules were aggregates they would possess no structural specificity, and although they would be roughly the same size and shape, would vary slightly according to the number of smaller molecules in them. If all the molecules were specific in structure, then those of a given type would all have exactly the same structure, size and shape. When these controversies began there was no way to test the theories experimentally to choose between them. The only methods available for studying macromolecules were those of colloidal† chemistry which did not have suciffient resolution to detect such small differences; it was not until 1926, when Svedberg developed the ultracentrifuge, that the necessary experiments could be carried out.

Svedberg's experiment

We have given you the two theories which were put to experimental test and give you (below) the information about how the test was performed and what the possible results were.

A sample of haemoglobin was placed on the top of a continuous sucrose density gradient (S100[1]) and subjected to a high centrifugal force for several hours. It was assumed that if aggregates were present they would not be broken up by this process. The molecules came to rest (i.e. in equilibrium between the upward buoyant forces and the downward gravitational forces) where the sucrose density was the same as their own. The two possible results are those shown in Figure 1.

specificity

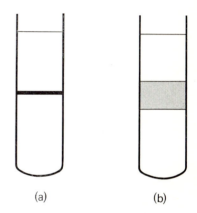

(a)   (b)

Figure 1 Two possible results of Svedberg experiment.

From what you know about centrifugation, decide which type of macromolecular structure would fit which result and why.

Tube (*a*) specific molecule theory.
Tube (*b*) aggregate theory.

The aggregates, differing slightly in their size and mass, and therefore densities, would come to equilibrium at different positions in the density gradient and hence produce a diffuse band. On the other hand, if all the molecules were of exactly the same size and mass they would all be at the same position in the gradient and form a sharp (i.e. narrow) band.

The result which Svedberg obtained was that of a narrow band, thus disproving the aggregate theory. His evidence, although not a *proof* of the specific molecule theory, was in agreement with its predictions.

Much later (1950s) X-ray crystallography (S100[2]) brought much more confirmatory evidence that at least some macromolecules (proteins) were specific in structure. Moreover it indicated that they were folded in a very precise manner. Although Pauling and Corey had again advanced the idea of the absolute specificity of macromolecules (namely, that a particular molecule had its own specific structure) it was not until 1954, when Sanger determined the amino-acid sequence of insulin, that biologists really accepted it, for proteins at least.

†*Terms marked with a dagger are defined in the Glossary on p. 42*

(When you come to look at the methods used to determine the amino-acid sequence of a protein (p. 15) you will realize the impossibility of the task if all the molecules were not exactly the same.) Sanger's success served to remove the despondency at the time about the possibility of *ever* defining the sequence of monomers (S100³) in a specific macromolecule. Since then, rapid progress has been made in the understanding of macromolecular structure and there is a tendency to forget, or at least not to appreciate, the very brief past upon which it rests.

We shall now look in more detail than you did in S100 at the structures of biological macromolecules to see how they are able to perform such a wide variety of functions with such accuracy.

1.1.1 The structural hierarchy

To facilitate the description of a macromolecule, the complete structure may be considered in terms of a hierarchy of increasing levels of complexity, which are named primary, secondary, tertiary and quaternary structures. The descriptions at these levels can be combined to give a statement of the complete structure, which ranges from the general shape and dimensions of the molecule (e.g. spherical with a diameter of 50 ångströms (Å)†), to the relative positions of the individual atoms.

At the present time such a description is possible for only a very few macro-molecules, all of them proteins. The complete structures of the nucleic acids are in most cases far from being solved, and as some of the polysaccharides seem not to possess specific structure in terms of *numbers* of sugar units, only a general structure can be defined for them. The case of the polysaccharides would appear to contradict what was previously said about specificity of structure, and, indeed, in terms of the *size* of the molecules (i.e. the number of monomers) this is true. However they do have specificity in their sequence of monomers (which are generally small repeating sequences). This lack of definite size is related to their functions as you will see later (p. 20). The reason why so little is known of macromolecular structures should become clearer as you read through this Unit.

We will start at the experimental beginning, the extraction and purification of the molecules, and examine the ways in which macromolecules can be obtained.

1.2 Separation of Macromolecules

Study Comment

Section 1.2 examines the problems of separating macromolecules from source material, tracing where it has gone and whether any purification has been achieved. Sections 1.2.1 and 1.2.2 look at two different criteria by which macromolecules can be separated, and Section 1.2.3 at how purity can be determined and what purity means. You should be able to recall the principles of the methods and their limitations, and to understand how the experimental data is interpreted.

Biological material is complex, and relatively susceptible to destruction or damage by extraction procedures. To remove the substance of interest from a chosen source, one must be rather careful not to expose it to extreme conditions which may alter it from its natural state, otherwise the results obtained after any analysis of it may well bear no relation to the material as it was *in vivo*. (Of course what constitutes an 'extreme condition' can only be proven by its deleterious effects on the substance, which implies knowledge of what the natural state is! Rather a vicious circle.)

Imagine that the substance one is interested in is the red pigment observed to be associated with the corpuscles in blood. There are three related problems involved in any attempt to obtain it:

(*a*) how to separate it from the corpuscles;

(*b*) how to decide that it is what was originally wanted;

(*c*) how to decide how pure it is.

The last two problems are usually two aspects of the same thing. Some feature of the original substance (called a 'marker') is used to enable one to follow it through different treatments. In this case it could be the red colour. This feature can also be used to determine whether or not the substance is getting purer. However, it is not just the *amount* of marker present which is measured, it is its *specific activity* (SA). This measurement of specific activity is of great importance in biochemical research because it enables a standardization of methods to be achieved by relating all experimental values to a common denominator. For example, if several experiments on the activity of an enzyme are being performed, maybe in different places and at different times, they can only be compared with one another if all are calculated in terms of the enzymic activity per milligram of protein. Just measuring the absolute amount of activity is no use, because this will vary depending upon how much starting material was there. Even if this could be made constant, losses would not necessarily be the same in each experiment and so the absolute activity could still fluctuate. Specific activity is a function of neither the starting nor the finishing amounts of the substance but of the *purity*. For those of you who are not sure of this, or don't believe it, we have given in Table I the results of two hypothetical experiments on one enzyme.

specific activity

Table 1 Two experiments to purify an enzyme

purification step	total enzyme activity (EU)	total protein (mg)	specific activity (EU/mg protein)
EXPERIMENT 1			
original extract	1 000	25.00	40
A	760	9.50	80
B	500	2.80	180
C	250	0.55	560
EXPERIMENT 2			
original extract	4 000	100.00	40
A	2 000	25.00	80
B	1 000	5.60	180
C	600	1.07	560

You will see that although the losses at each step are not constant, the degree of purification (i.e. the increase in SA) is, and is dependent not on enzyme activity or on quantity of protein material alone, but on both of these values.

But what if the molecule is not an enzyme, as it may well not be in the red pigment above? Red colour is not an enzymic activity, but it can be determined nevertheless. In your Home Experiment in S100 you measured the intensity of colours; that could be done here too. If the molecule were the substrate for an enzyme it might be possible to measure it by the ability of a known amount of enzyme to degrade it to some particular products. Alternatively, and of increasing importance at present, it might be possible to 'label' it with a radioisotope (S100[4]) and follow the amount of radioactivity. (You will see in TV programme 4 how this can be done.) If a substance has been labelled in this way the specific activity of the radioactive material is commonly called its specific radioactivity.

The problems of 'whether it is what was wanted' and 'how pure it is' have been solved relatively simply in this chosen case, by reference to the red colour. But what of the problem of separation? All separation techniques depend upon differences existing between the substance required and any unwanted material. These are exploited as a means of removing the substance from all or most of its contaminants. The more contaminants removed in any one step, the finer its discrimination. Obviously one could stop the purification process at any point, depending upon the degree of purity required, but to obtain macromolecules of high purity, say 95 per cent, one usually has to use a series of methods. It is easy to refer glibly to 95 per cent purity, but what does it mean? To state that a substance is 95 per cent pure assumes a knowledge of its features when it is 100 per cent pure, as a reference point. The whole idea hangs upon an *operational* definition of purity: that is, if no increase in purity can be produced or contaminants detected by the application of all conceivable methods, then the substance *is* 100 per cent pure! (See p. 13.)

The methods used to remove contaminants, and hence to increase the purity, separate the components of the mixture on the basis of various criteria, e.g. size, charge and solubility. Any contaminants which resemble the required substance in one respect can be separated from it by another in which they are quite different. The first step in separation is usually to extract a whole class of macromolecules, e.g. proteins, in which the substance is found to reside. Most of this material will be unwanted and the next step is to successively fractionate the selected class until only the desired substance is left. There are various methods available for doing this.

1.2.1   Segregation according to size

Segregation according to size can be applied to all classes of macromolecules. It can be carried out in essentially two ways; either by centrifuging in a density gradient (as for example in the Svedberg experiment) or by 'molecular sieving' through a material which has fine pores in it. The former method you met in S100, Unit 14 (albeit for cellular organelles) and if you don't remember the principles you should now re-read it quickly. The latter method, commonly called 'gel filtration', is of great importance in molecular studies and has been refined over the past few years into a technique of great sensitivity, especially since the replacement of natural porous earths by synthetic molecular sieves. A slurry is made from a solid which itself consists of a large number of small particles, each of which has holes or pores running through it.

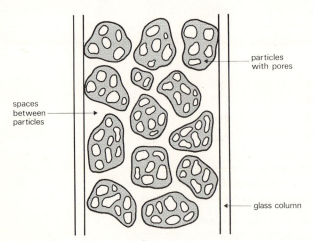

Figure 2   Gel filtration.

In the synthetic materials these pores are obtained by controlled cross-linking (S100[5]) of polysaccharide polymers; the greater the degree of cross-linking, the smaller the pores. In the slurried state the spaces *between* the particles are much larger than the pores, so that if the slurry is placed in a column, i.e. a long hollow cylinder, and a mixture of macromolecules with varied dimensions washed through it (eluted) the molecules appear separately at the bottom. The largest molecules appear first and the smallest last. Can you think of the reason for this?

The space *within* the particles, i.e. the pores, is many times greater than the space around them. As a consequence small molecules have a much greater volume in which to disperse within any given small band (horizontal section, c.f. centrifugation) of gel or slurry than have the large molecules. The latter, therefore, are washed through the gel very rapidly. In fact, they will appear at the bottom of the gel when one 'void volume' of eluant has been added. (A void volume is equal to the volume of spaces between the particles.) The smallest molecules will only appear when a volume of eluant equal to the volume of the spaces between *and* inside the particles has been added to the column. This may indicate to you another possible use for gel filtration; it can be used to separate any unwanted inorganic ions from the macromolecular material and can be used in place of dialysis† (see TV programme 2) to 'de-salt' preparations.

The eluant is collected as small sequential samples (fractions) from the bottom of the column; the feature or marker being used to follow the wanted substance, e.g. the red colour in our previous example, is then determined in each fraction. The level of protein or nucleic acid present might also be determined for each sample (to derive specific activities and find how well other substances had been separated) and graphs of the type shown in Figure 3 obtained.

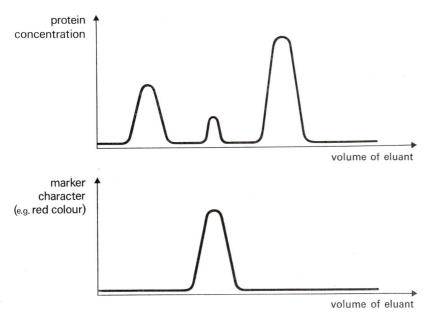

Figure 3   Eluants from gel filtration monitored for marker and protein.

From these graphs it can be seen that the material which contained the marker was eluted from the column in quite a small volume of eluant compared to the total which was passed through. It was not dispersed throughout the whole eluant and appeared to be separated quite clearly from the other substances present. In TV programme 2 you will see more uses of gels in columns and how the eluant can be monitored.

However, it is not always as easy as our example suggests to obtain a good separation by this method. Two substances may be very alike in size and run off the column so close together that no separation can be achieved. Some other method must be used for such cases which is not dependent upon size; this could be their difference in electrical charge.

1.2.2   Segregation according to charge

Proteins are eminently suited to methods which utilize charge effects, because of their varied composition of amino acids, some of which have charged side-groups. If you look at Figures 9–11 and 9–13, pp. 187–8 in *Cell Structure and Function*, you will see, respectively, the effects of pH upon the charges carried by amino acids and the charges which side-groups can carry. Don't concern yourself with the pK values; they will be dealt with for proteins as a whole later in this Unit (p. 33). A *zwitterion* is a molecule which carries both a positive and negative charge at the same time, e.g. the centre structure of Figure 9–11.

11

In a large protein with several hundreds of amino acids, the net, or overall charge on it will be the sum of all the positive and negative charges on the side groups (plus, of course, the charges on the carboxyl and amino groups at each end). Whether a side-group is charged or not depends partly upon the pH of the solvent (named the 'medium') and partly upon which groups are near it in the molecule.

> **ITQ 1**   What would happen if an electric field was applied across a solution containing several different proteins?
>
> *Answer on p. 38.*

This operation is called *electrophoresis*. It can be carried out in solution but is normally done on an absorbent paper or gel (either starch or a synthetic polymer called polyacrylamide). The paper or gel is saturated with a liquid of suitable pH (found by trial-and-error) and the mixture of proteins added to it. A voltage is then applied across the paper or gel for a chosen length of time (S100[6]). (See Fig. 9–2 in *Cell Structure and Function* and TV programme 3.) **electrophoresis**

Unless the separated components are coloured, some means of detection is needed. This could be done either by staining the paper or gel with a dye which selectively binds to proteins or by looking at it under light (most often ultra-violet light) at a wavelength at which the substances absorb. The gel or paper can then be cut up and the separated components dissolved out.

Another way of obtaining separation by charge is to use a long glass or plastic tube (a column) held vertically and filled with a synthetic resin which has charged groups on it. The arrangement is similar to that used for gel filtration (see p. 10).

A resin with positively charged groups will bind all negatively charged molecules, the strength of binding being greater the more negatively charged the molecule. All positively charged molecules will not be bound and can be washed out. By gradually lowering the pH of the eluant, the bound molecules will be removed from the column; those with the smallest negative charge coming off first and those with the greatest last. Thus a separation by charge can be affected. The method works by a shift in the equilibrium position through the addition of hydrogen ions, i.e. a lowering of pH (S100[7]). **ion exchange resin**

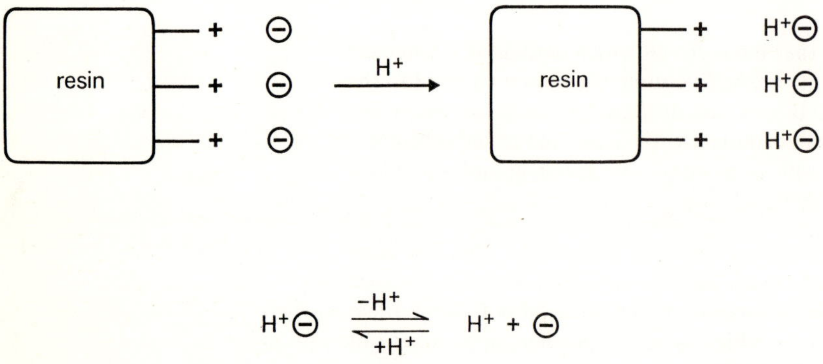

Figure 4   *Equilibria between charged molecule and charged resin.*

The less negatively charged a molecule is, the fewer hydrogen ions are required to neutralize it and to free it from the resin. Similarly a separation of positively charged molecules can be obtained by using a negatively charged resin and successively increasing the pH of the eluant.

As we said earlier, these charge separations are very useful for proteins; they are less so for nucleic acids and polysaccharides which have very little heterogeneity of charge and are usually separated by size differences, although their differences in composition (i.e. sequence of monomers) can be exploited to some extent by their differing abilities to bind to certain synthetic resins. We will not go into this here.

There are a few special methods for the separation of proteins which are dependent upon their solubilities. The solubility of any protein varies with the ionic strength (i.e. concentration of ions) of the solvent. By altering this variable, specific proteins can be precipitated out from solution at their points of lowest solubility, and removed by centrifugation. Ammonium sulphate is most commonly used for this purpose (TV programme 1 and Home Experiment). The technique is widely used for proteins of a molecular weight (MW) less than 150 000. Above this value, the molecules are usually composed of several sub-units (see p. 30) and ammonium sulphate has a tendency to cause them to dissociate. An alternative method is based upon the feature mentioned earlier (p. 11), namely that pH changes cause an alteration in charge on proteins, and at certain pHs proteins bear no net charge. This 'point-of-no-charge' is called the *iso-electric point* of the protein and is its point of lowest solubility. Thus by varying the pH of the medium, proteins can be precipitated separately at their individual iso-electric points. (You will be using this technique in your Home Experiment.)

iso-electric point

These are just the principles involved in separating macromolecules. In practice, a great deal of trial-and-error is needed to find a series of methods which yield the purest product possible, as you will certainly find in your Home Experiment. The pH and ionic strength of the media can drastically affect the results, so not only must several different methods be tried, but also the conditions under which these are performed must be varied. When a point is reached at which the specific activity of a product cannot be increased any further by any variations, the product can then be tested for purity (see below and TV programmes 1 to 3).

1.2.3 Criteria of purity

If only one substance is present in a macromolecular preparation, what will be observed if it is subjected to, say, electrophoresis or gradient centrifugation under several conditions?

There will be only one band.

If it is completely (100 per cent) pure, the use of several criteria of purity will all indicate that only one type of molecule is present. Some of the techniques which are used to *separate* macromolecules can be used as criteria in this way, usually under several conditions, e.g. electrophoresis, gel filtration or ultra-centrifugation. Another very sensitive method of monitoring the presence of impurities is to determine the solubility of the substance. A typical result from such an experiment is shown in Figure 9–5 of *Cell Structure and Function*, p. 180.

You should now attempt SAQ 6.

1.3 Structure Determination – the First Steps

Study Comment

Section 1.3 examines the methods which were and are used for measuring the gross characteristics, such as size, of macromolecules, in particular ultra-centrifugation. You should be able to distinguish between the principles upon which the two methods of centrifugation are based and point out how wrong information can sometimes be obtained.

Once a pure preparation of a macromolecule has been obtained, the determination of its structure can begin. The gross characteristics, such as size and molecular weight, are generally the first to be measured and there are several ways of doing this.

One of the oldest approaches, which dates back to the early days of organic chemistry, is to measure the relative proportions of all the elements present and derive an empirical minimum formula, and hence a minimum molecular weight. Haemoglobin, for example, has a minimum formula of $C_{712}H_{1130}N_{214}S_2O_{245}Fe$ and therefore a minimum molecular weight of 16 700. However, for a molecule composed of several subunits (see Section 1.7) either similar or identical, the answer obtained by this method will be only a fraction of the true answer. In the case of a molecule composed of four identical subunits, the MW derived by this method will be one quarter of the actual MW. How then can the true structure be obtained?

The most recently developed technique, almost the complete antithesis of the earlier chemical methods, is electron microscopy. With this method the actual size and shape of the molecule can be measured and from these its MW calculated with a knowledge of its density (which is determined separately). Unfortunately, the resolving power of the electron microscope is a limiting factor in determining how small an object can be seen. Only molecules of molecular weights of several million and above can be dealt with in this way. Two electron micrographs (that is photographs from an electron microscope) are shown in Figure 9–10, p. 184, in *Cell Structure and Function* with a brief explanation of how they are obtained.

Before much confidence can be placed in the molecular weight derived for a macromolecule, it must be obtained by several different methods with a fairly close agreement between them. Some techniques are more susceptible to errors through dissociation or association (S100[8]) than others, and widely differing results can be obtained. (Look at Table 9–3, p. 185, in *Cell Structure and Function*. Don't concern yourself with the actual techniques used but note the discrepancy in the insulin MWs as compared to those of the other macromolecules.) The ultracentrifuge yields two ways of determining molecular weights and at the same time gives information on size and shape. These two methods are called *sedimentation equilibrium* and *sedimentation velocity*. In the former, a continuous density gradient (of sucrose or, more usually, caesium chloride) is spun in the ultracentrifuge and the molecules form a sharp band in the gradient (S100[1] and p. 7 this Unit). The position at which the molecules reach equilibrium is dependent upon their density and molecular weight. It is possible to measure density by an independent method and so the molecular weight can be obtained. To avoid diffusion of the molecules the equilibrium must be reached slowly, which may take days or even weeks.

sedimentation equilibrium

The sedimentation velocity method, which is much faster than the equilibrium method, measures the velocity at which the band of molecules moves down through a uniform density gradient when centrifuged at high speeds. This measurement is usually done by means of an optical system built into the centrifuge, which can photograph the contents of the tube at various times during the 'run'. (There is an example of this kind of photograph in Fig. 9–3, p. 178 in *Cell Structure and Function*.) The shapes of macromolecules can very often be inferred from centrifugation data, in that sedimentation characteristics differ between differently shaped molecules.

sedimentation velocity

ITQ 2   There is another method (which you met earlier in this Unit) which can be used to determine the MW of a macromolecule; do you know which it is?

All the techniques which you have looked at so far are, in a sense, a preamble, albeit a very necessary one, to the work of structure determination. For a great many macromolecules, this is as far as research has proceeded, because the problems of purification have not been sufficiently resolved or the next steps are too difficult. However, now that you can see in principle how to isolate a macromolecule and determine its purity, we can go on to examine the methods used to elucidate its structure. We will proceed through the levels of complexity from primary to quaternary structures, looking at some of the ways in which structures can be determined at each of those levels, and at something of what is known about them today.

1.4 Primary Structure

Study Comment

Section 1.4 is large and deals with the different types of primary sequences which exist in three classes of macromolecule, with some of the experimental methods for determining them. The three classes are considered separately: proteins (in Section 1.4.1), nucleic acids (in Section 1.4.2) and polysaccharides (in Section 1.4.3). Much of the primary structure of polysaccharides was covered in S100, Unit 13, and in *The Chemistry of Life;* which you will be directed to re-read. You should be able to recall which types of small molecules are the monomers for each class of macromolecule, how they are bonded together and what methods can be used (although not with their chemical reactions) to determine these sequences. You should also be able to compare and contrast the similarities and differences between these structures, the methods used and the results obtained for all three classes. Your ability to deduce structures from experimental data is tested in the SAQs.

In *The Chemistry of Life* and in S100, Unit 13 you met briefly the ideas involved in the determination of the primary structures of macromolecules. In all such operations the primary sequence, that is the *order* in which the monomers are linked together, is deduced from data derived from the controlled degradation of the large molecule into smaller pieces which are susceptible to detailed analysis. The specific technique of degradation used depends upon the type of macro-molecule being studied. Here we shall look at three types, proteins, nucleic acids and polysaccharides, to see where the similarities and the differences lie.

1.4.1 Proteins

Proteins are made up of a large number of amino acids, of which there are twenty different kinds. The relative amounts of each amino acid in a particular protein molecule can be measured by complete hydrolysis (S100[9]) of the molecule and analysis of the amino acids so set free. This can be done automatically with a machine called (logically!) an amino-acid analyser. The actual sequence in which these were originally joined together is determined by use of enzymes and chemical reagents. The enzymes are called *peptidases* and are capable of cleaving polypeptide chains (by catalysing the hydrolysis of peptide bonds (S100[10] and *Cell Structure and Function*, Figure 9–16, p. 191)) at specific points, either in the interior or at the ends. These are called endopeptidases and exopeptidases respectively. For example, chymotrypsin cleaves the chain on the carboxyl side of every phenylalanine (Phe), tyrosine (Tyr) and tryptophan (Try) when these amino acids are not at the end of the chain. Chymotrypsin is therefore an *endo-peptidase.*

peptidases

endopeptidase

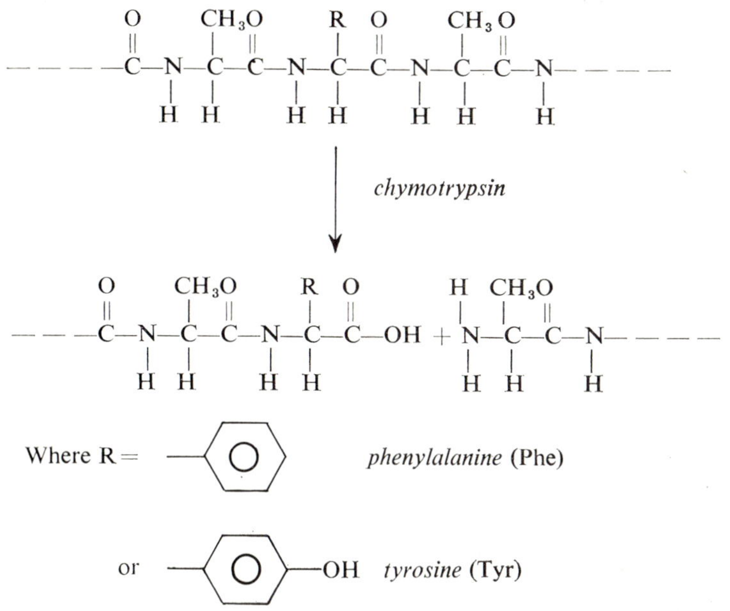

Figure 5   Action of chymotrypsin.

15

The group of enzymes called aminopeptidases cleave off the N-terminal amino acids (*The Chemistry of Life*, p. 57), as shown in Figure 6,

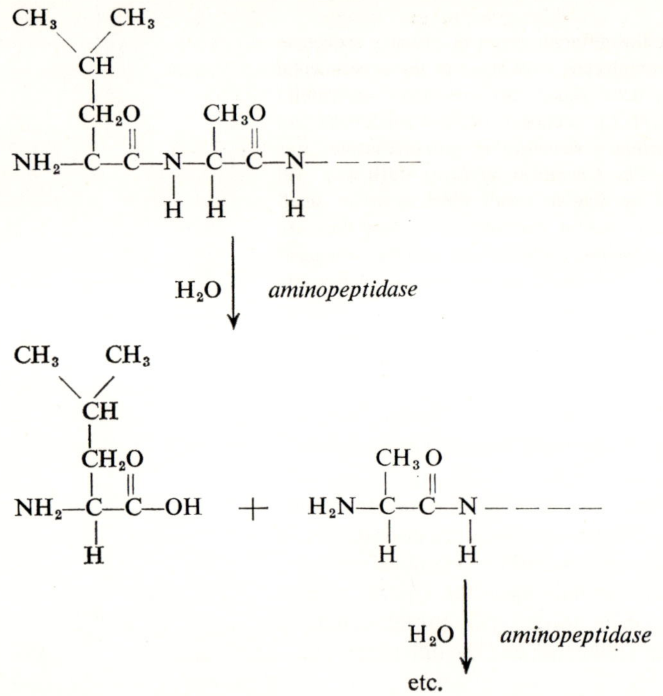

Figure 6   Action of aminopeptidases.

and are therefore *exopeptidases*. (The rate of cleavage is dependent upon the identity of the N-terminal amino acid; not all are cut off at the same rate.) Another group of exopeptidases is the carboxypeptidases, which, as their name suggests, cut off C-terminal amino acids.

exopeptidase

Other enzymes have different specificities, so that by using one enzyme to reduce the protein to smaller fragments, separating them (perhaps by one of the methods discussed above for protein purification) and then treating these with a different enzyme, a large number of small polypeptides can be obtained. An ideal size for the final fragments is about four amino acids; these fragments can also be separated from one another and the sequences of the amino acids in them determined.

> **ITQ 3**   In Figure 7, we have given you the outlines of an enzymic hydrolysis of a protein. From it, what can you deduce about:
>
> (*a*) the sites of trypsin-catalysed hydrolysis of peptide bonds?
>
> (*b*) the general positions of peptide bonds cleaved by trypsin?
>
> (Look at Fig. 9–13, p. 188, in *Cell Structure and Function* if necessary.)

NH$_2$·Ser·Ala·Arg·Ala·Ala·Phe·Asp·Pro·Lys·Thr·Lys·Ala·Glu·Pro·Val·COOH

$$H_2O \quad \downarrow \quad chymotrypsin$$

NH$_2$·Ser·Ala·Arg·Ala·Ala·Phe·COOH + NH$_2$·Asp·Pro·Lys·Thr·Lys·Ala·Glu·Pro·Val·COOH

$$H_2O \quad \downarrow \quad trypsin$$

NH$_2$·Ser·Ala·Arg·COOH + NH$_2$·Ala·Ala·Phe·COOH + NH$_2$·Asp·Pro·Lys·COOH + NH$_2$·Thr·Lys·COOH + NH$_2$·Ala·Glu·Pro·Val·COOH

Figure 7   Hydrolysis of protein by trypsin and chymotrypsin.

The determination of the sequences of the small fragments is begun by finding which are the C- and N-terminal amino acids. This is done by reacting the free amino (NH$_2$–) and carboxyl (—COOH) groups with reagents which give easily

N-terminus

16

detected derivatives. Amino groups will react with dinitrofluorobenzene (DNFB) to give an amino-acid derivative which is yellow, a dinitrophenyl amino acid (DNP-aa). (This method can also be used to find out which is the N-terminal amino acid in the intact protein.) Although the side-group amino groups of such amino acids as lysine will also react, only one α-DNP-aa will be produced; that from the N-terminus. (See Fig. 8.)

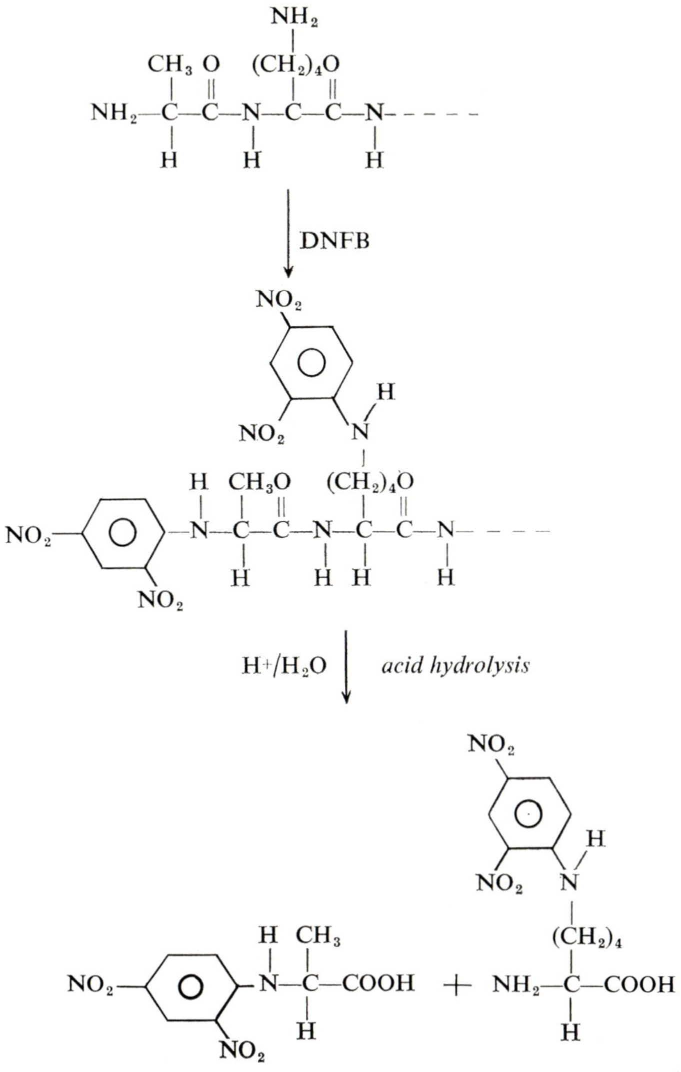

Figure 8 Reaction of DNFB with N-terminal and side-group amino groups.

ITQ 4 Using common sense only, can you think of a case where no α-DNP-aa can be found?

The C-terminal amino acids can be treated in an analogous fashion, although C-terminus the methods used are far from easy and we will not deal with them here. In Figure 9 we have given you a schematic example of how part of the structure of a small polypeptide can be deduced using DNFB and a specific enzyme.

$$NH_2 \cdot aa_1 \cdot aa_2 \cdot aa_3 \cdot aa_4 \cdot aa_5 \cdot aa_6 \cdot aa_7 \cdot COOH$$

enzyme E cleaves at COOH side of aa_3

$$NH_2 \cdot aa_1 \cdot aa_2 \cdot aa_3 \cdot COOH + NH_2 \cdot aa_4 \cdot aa_5 \cdot aa_6 \cdot aa_7 \cdot COOH$$

DNFB followed by acid hydrolysis

$$NH_2 \cdot aa_1 \cdot COOH + NH_2 \cdot aa_2 \cdot COOH + NH_2 \cdot aa_3 \cdot COOH$$

Figure 9 Determination of a tripeptide sequence with DNFB and an enzyme.

In this way the sequences of all the small fragments of a protein can be determined, and these can then be fitted together to give the sequence of amino acids in the whole protein. Generally it is necessary to perform the whole operation several times using different enzymes and 'controlled acid hydrolysis' (that is with different strengths of acid for different lengths of times) to obtain enough overlapping fragments. As we said on p. 7, this was first done with one of the chains of insulin which contains 30 amino acids, but it still took ten years to perform! Of course, at the start of this period none of the techniques mentioned above were available. Sanger and his associates had to develop them for themselves. If you look at Figure 9–19, p. 196–7, in *Cell Structure and Function* you will see the fragments which they obtained and how they were fitted together. (Do not try to remember the cleavage points of the enzymes or any of the fragments.)

If a protein has one or more pairs of cysteine amino acids (Cys), these may be joined together to form disulphide bridges (see Fig. 10).

$$NH_2 \cdot aa_1 \cdot aa_2 \cdot aa_3 \cdot aa_4 \cdot aa_5 \cdot aa_6$$

Figure 10   A polypeptide with a disulphide bridge.

Before such a protein can be sequenced, these disulphide bonds must be broken. If this were not done, pieces would be produced made up from two fragments joined together through the disulphide bonds. These bonded pieces would then have two N-termini and two C-termini, as shown for example in Figure 11.

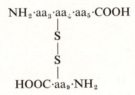

$$NH_2 \cdot aa_3 \cdot aa_4 \cdot aa_5 \cdot COOH$$

Figure 11   Fragment from degradation of polypeptide in Figure 10.

However, if the disulphide bonds are cleaved before the protein is degraded, then if more than one pair of cysteines are present it will be impossible to tell which pairs were originally connected. To overcome this problem, when the structure is known, the molecule can be degraded *without* breaking the disulphide bonds, and the pairs of cysteines identified by the amino acids around them. More will be said about these important amino acids later in this Unit.

At the time of writing, the sequences for about 500 proteins have been completely worked out. This total includes variants of particular molecules between species, e.g. horse and human haemoglobin, which are counted as two proteins and not one (S100[11]).

1.4.2   Nucleic acids

In determining the primary structures of nucleic acids a much more difficult situation arises than with proteins. There are only four different monomers (the nucleotides) in each type of nucleic acid as compared with twenty (the amino acids) in proteins. These are, in DNA, adenine (A), cytosine (C), guanine (G) and thymine (T) and in RNA adenine (A), cytosine (C), guanine (G) and uracil (U) (S100[12]). Consequently, the fragments obtained from the degradation of nucleic acids are very similar to each other. This problem is exacerbated in general by the relative lack of specificity of the enzymes (DNAases and RNAases) available, compared with those available for proteins. They can only distinguish between types of bases, i.e. purines or pyrimidines, rather than between the bases themselves. The presence of several 'unusual' bases in transfer

RNA (S100[13]) has eased this problem to some extent and some tRNA sequences have been completely worked out. However, for most nucleic acids very little is known.

Look at the results of a typical degradation of a short segment from DNA, shown in Figure 12.

pA·A·G·C·G·A·T·T·A·T·G·C·T·A·G·G·C·T·A·A·A·T·C·C·C·OH

> enzyme cleaves on deoxyribose side of
> phosphate bond of purines

A, A, G, C·T, A, T·T·A, T·G, C·T·A, G, G, C·T·A, A, A, T·C·C·C
i.e., 5A; 3G; 1 C·G; 1 T·G; 1 T·T·A; 2 C·T·A; 1 T·C·C·C.

Figure 12 Degradation of DNA. (All phosphate and free hydroxyl groups are omitted from fragments.)

The positions of the quite large numbers of repeating sequences of similar nucleotides cannot be stated with any accuracy. Even though the amino-acid sequence of a protein may be known, it is not possible to deduce from this the sequence of nucleotides in the DNA or RNA on which it was formed by reference to the genetic code. In S100, Unit 17, you learned that each amino acid in a protein is coded by a sequence of nucleotide bases in the DNA and hence in the mRNA. However, because most amino acids have more than one trinucleo-tide codon (i.e. redundancy) it is not possible to work back from protein to nucleic acid.

1.4.3 Sequencing of transfer RNA

The one kind of nucleic acid which has been successfully tackled is tRNA; this is easier to sequence because of several factors. The first is the size of tRNA molecules; they are small, each consisting of about 80 nucleotides, as compared to a typical mRNA with about 450 nucleotides (as for example that for myoglobin). Secondly, although there are about 60 types of tRNA in yeast for example, they can be separated fairly readily on a big enough scale to allow further analytical work to be performed. (This point – the scale of isolation – is one which often arises in biochemistry – sometimes several *tons* of a tissue are used to obtain a few mg of material!) The separation is achieved by making use of the function of tRNA in the cell, that of transferring amino acids (S100[13]). One tRNA 'loaded' with its appropriate amino acid differs in its solubility properties from other tRNAs loaded with their amino acids; the separation method uses these solubility differences. Take as an example two substances, X and Y, which differ in that X is three times more soluble in a liquid A than in another liquid B, while Y is three times more soluble in B than A. If the liquids A and B are immiscible (do not mix), then when X and Y are dissolved in equal quantities of A and B and allowed to equilibrate, the result will be as shown in Figure 13.

If the A layer is removed, shaken with a fresh equal quantity of B and allowed to equilibrate then the result will be as shown in Figure 14.

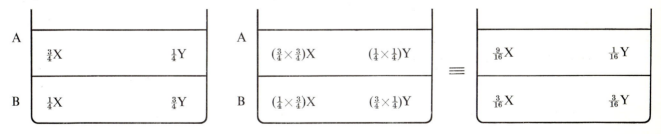

Figure 13 First equilibration of two solutes between two solvents.

Figure 14 Second equilibration of solutes with fresh solvent B.

If X and Y were originally present in equal amounts, the ratio of the amounts of X:Y in the A layer will have risen from 1:1 to 9:1. Obviously some X is being lost each time into the B layer, but the *purity* is increasing, and although the point of theoretical absolute purity can never be reached, in practice a satisfactory result can be achieved. The trick of this method is to have as widely differing solubilities as possible by choosing the liquids carefully.

With tRNAs this process must be repeated thousands of times in a mechanical mixer-separator. Usually the two liquids move through the machine in opposite directions and the process is called 'counter-current distribution'. The first tRNA to be extracted in this way was $tRNA_{Ala}$.Ala (that is the tRNA specific for alanine with its alanine attached) in 1966; but since then several others have been obtained.

The next problem is to degrade the molecules into manageable fragments and separate these from one another. Very careful treatment with an RNA hydrolysing enzyme (an RNAase) at 0 °C splits the molecule into two almost equal parts which can be separated by chromatography (see TV programme 3). A slightly more vigorous treatment gives a few more fragments and so on until a whole range of fragments from 1 or 2 nucleotides up to 40 has been obtained. These can be analysed for their composition by complete alkaline hydrolysis (i.e. hydrolysis into single nucleotides with alkali) followed by paper chromatography to separate and identify the nucleotides. The use of several different enzymes under a variety of strictly controlled conditions yields enough overlapping fragments for the whole sequence to be deduced. This process is greatly simplified by the presence in tRNAs of several 'unusual' bases, e.g. inosinic acid (see Fig. 15), which form landmarks.

unusual bases

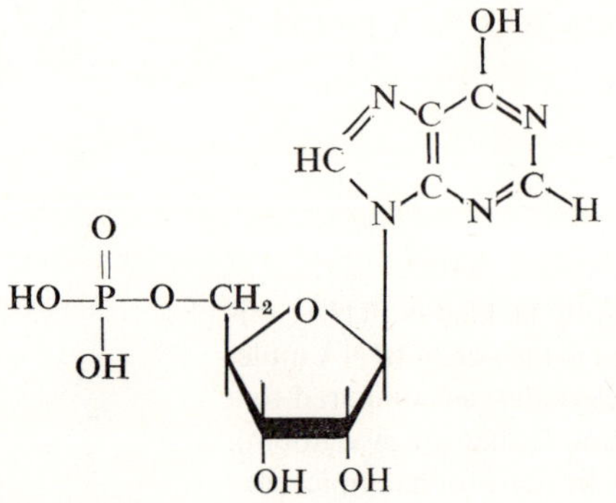

Figure 15 Structure of inosinic acid.

The complete sequence of $tRNA_{Ala}$ is shown in Figure 16. Note the occurrence of these unusual bases and the presence of the anticodon and the amino-acid binding site (S100[13]).

```
        Me                     DiH     DiH        DiMe                        Me
        |                      |       |          |                           |
p G-G-G-C-G-U-G-U-G-G-C-G-C-G-U-A-G-U-C-G-G-U-A-G-C-G-C-G-C-U-C-C-C-U-U-I-G-C-I-ψ-G-G-G-A-G-A-G-U-C-U-C-C-G-
                                                                    anticodon
G-T-ψ-C-G-A-U-U-C-C-G-G-A-C-U-C-G-U-C-C-A-C-C-AOH
                                             ↑
                                        amino acid
                                        binding site
```

The unusual bases are:

MeG – 1-methyl guanine
DiH – 5,6-dihydrouracil
DiMeG – dimethyl guanine
MeI – 1-methyl inosine

I – inosine
ψ – pseudouracil
T – ribothymine

Figure 16 Sequence of bases in tRNA.

1.4.4 Polysaccharides

S100[14] and *The Chemistry of Life*, pp. 45–52, both of which you should now read, deal more than adequately with most of the aspects of polysaccharide primary structure which you need to consider at this stage. The following notes expand certain points and summarize the main facts of which you need to be aware.

(*a*) The relatively small variety of monomers (sugar molecules like glucose, galactose) makes sequencing easy for most polysaccharides because they appear in short *repeating* sequences. (This is the opposite of the case of the nucleic acids.) Branching points are detected by means of enzymes which have specificities for either 1–4 or 1–6 but not both.

(*b*) Note the differences and similarities between glycogen, starch and cellulose in terms of both their structures and function.

(*c*) Even in the case of a homopolysaccharide, several structures can be produced by virtue of the different bond conformations and types, e.g. α and β, 1–4 and 1–6.

(*d*) Note how large most of the polysaccharides are, e.g. glycogen with a MW of several millions.

(*e*) The functions of the molecules are directly related to their structures. Cellulose is a linear molecule which can form fibrous structures. Glycogen is globular, but more importantly has no fixed size. It is an 'add-on/take-off' type of molecule which serves as a glucose store in animals. During times of plentiful food, glucose molecules are enzymically added on to the ends of the free polymer chains by 1–4 or 1–6 linkages whereas, in starvation conditions, they are enzymically removed to maintain the blood sugar level. This random addition and subtraction generates glycogen molecules of varying sizes.

(*f*) Polysaccharides have many functions and structures other than those dealt with in these texts. The glycoproteins which are part polysaccharide and part protein form the rigid material of bacterial cell walls (see diagram on p. 121 of *Cell Structure and Function*). Other polysaccharides are formed in brain tissue and are important in clinical diagnoses of diseases.

1.4.5 Summary

You have seen that, for all biological polymers, the primary structures are determined in essentially the same manner, by breaking them into smaller pieces which can be analysed by a combination of chemical and enzymic methods. The differences between the methods applied and the progress made so far is dependent upon the class of molecule being studied. Perhaps, when more is known about the enzymes or the very specific chemical reagents which cleave macromolecules, it may be possible to 'tailor' them to be even more specific in their actions and so solve some of the problems which still exist.

1.5 Secondary Structure

Study Comment

Section 1.5 examines some of the secondary structures which have been found in various types of macromolecule, without reference to the ways in which they were deduced. Section 1.5.4 is an outline of some of the experimental procedures. All the structures mentioned are dealt with in relation to examples and, where possible, the relation between the structure and the function of a molecule is pointed out. Note that in the cases of keratin, silk, collagen and DNA the discussion of structure and function inevitably makes a jump forward to quaternary and higher orders of structure, that is to structures composed of more than a single molecule, as in a fibre. The chosen examples, therefore, are unfortunate in that respect but are nevertheless good cases in which to demonstrate the *kinds* of arrangements which are called secondary structure. You should be able to define secondary structure and give examples for different classes of macromolecule, with a very brief outline of the properties upon which their detection and analysis depend.

1.5.1 Nucleic acids

It might seem reasonable to assume that before an attempt at secondary structure determination can be made, the primary structure must have been worked out; in fact this is not so. The helical conformations of the nucleic acids were known long before many of the details of the primary structures, e.g. the DNA double helix. We will use here a loose definition of secondary structure as 'those regular conformations which result from hydrogen bonding (S100[15])'. As you read in S100, Unit 17, the DNA molecule is held together in a double helix partly by hydrogen bonds between pairs of purine and pyrimidine bases (see Figs 8.7 and 8.8, pp. 142 and 143, in *Cell Structure and Function*). The other forces which are involved are apolar interactions between the bases. (Apolar bonding in proteins will be dealt with in Section 1.6 of the Unit.) This base pair hydrogen bonding is also thought to occur in mRNAs and tRNAs (see Figs 8.23 and 8.26, pp. 161 and 167, in *Cell Structure and Function*). In these cases, the absence of total complementarity of the bases (i.e. not all the bases have a complementary base with which to pair (S100[16])) only permits short lengths of the molecules to form helices by hydrogen bonding.

The presence of two complementary strands in DNA allows a double helix to be formed by hydrogen bonding between every pair of bases. This is of considerable importance because, although the strength of each individual bond is low (bond breaking energy of approximately 13 kJ/mol (S100[17])), the stability of the molecule is quite high due to the summation of tens of thousands of hydrogen bonds. (The apolar forces between the bases have the same effect.) DNA serves the cellular function of containing the information needed to control metabolism (Unit 6 of this Course) and to direct the cell's growth and specialization; hence it must be relatively immune to degradation. At the same time it must be able to participate in some reactions, otherwise the information which it contains would be useless, and it also would not be able to replicate at cell division (S100[18]). The bases which code for the proteins are embedded inside the helix and are not very accessible. Here the advantage of summation of many weak bonds can be seen. If the hydrogen bonds are broken individually in succession along the length of the molecule, the two strands can be pulled apart. However, because the hydrogen (and apolar) bonds form spontaneously under the conditions in the cell, no external mechanism is needed to reform them in the correct order afterwards.

By analogy with DNA, mRNA and tRNA might be expected to exhibit some degree of hydrogen bonding and helices. Of the two, tRNA has been the most studied because of its size and, in some cases, known sequences. That some helical regions are present is generally accepted but where they occur and what overall shape they give the molecule is not known. Because the anticodon must interact with its complementary section (codon) on the mRNA, it is reasonable to assume that it is fairly exposed, on the 'surface' of the molecule. Similar considerations apply to the amino acid binding site. The rest of the molecule must interact with the small subunit of the ribosome in some precise manner during protein synthesis (S100[19]), but quite how this occurs is still a subject of speculation. A variety of possible conformations have been proposed since the sequence of tRNA$_{Ala}$ was worked out by Holley *et al.* in 1966 but, although they have gone into and out of favour, no one has yet been finally established. A possible structure is shown in Figure 17, with just the anticodon and amino-acid binding site marked.

1.5.2 Proteins

Protein secondary structure can be divided into two types, helical and sheet formations. In both cases the hydrogen bonding which produces the structure is between the C=O and the N—H groups of the peptide bonds. The α-*helix* (only one of a few possible helices) is the most common helix and is formed by hydrogen bonding between C=O of one amino acid and the N—H of another, three 'residues' (i.e. amino acids linked by peptide bonds) nearer the C-terminus in the chain. This is shown schematically in Figure 9–27, p. 204 in *Cell Structure and Function*. All amino acids except glycine (Gly) have two stereoisomers (S100[20]) called D and L[†] forms, of which the naturally occurring form is L. L-amino acids form a right-handed helix (when viewed with N-terminus at the top) in preference to a left-handed one, whereas the converse is true for D-amino acids (Fig 18). The right-handed helix formed by L-amino acids is much more stable than the left-handed one because of the 'steric hindrance' (repulsive forces between atoms when they are brought too close together (S100[21])) of the C=O

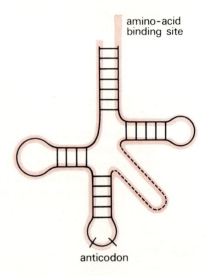

hydrogen bonds

Figure 17 A possible structure of tRNA.

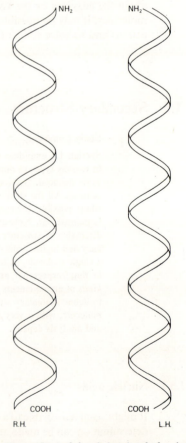

Figure 18 Left-handed and right-handed protein helices.

groups by the side groups of the amino acids. The reverse situation holds true for the D-amino acids.

The case of polyglycine is entirely different.

ITQ 5 Glycine has no isomers. Do you know why? (If you don't remember the structure look at Fig. 9–13, p. 188 in *Cell Structure and Function*).

The α-helix is a moderately stable structure in its own right, all the C=O and N—H groups forming hydrogen bonds with each other, although studies of the helix in solution have shown that water can compete effectively in this hydrogen bonding process. In the whole protein, therefore, other stabilizing forces must be present, and as the side groups of the amino acids all project out from the helix, they are able to fulfil this role.

The α-helix is usually found only in small quantities in most proteins, especially those of the globular type. Fibrous proteins often have much more. For example, hair, bone and horn are made from protein molecules with completely helical structures. Hair is composed of keratin molecules. Three of these α-helical molecules are wound around each other to produce a 'superhelix' called a 'protofibre'. Eleven protofibres are arranged in a (9+2) formation to form a 'microfibril' (see Fig. 19).

The (9+2) arrangement is not an uncommon feature in biological structures, being found in sperm tails and the cilia of some micro-organisms (S100 TV 18 and *The Microstructure of Cells*,* p. 30). Hundreds of microfibrils are embedded in an amorphous† protein matrix, which has a variable content of sulphur-containing amino acids, i.e. cysteine, producing a 'macrofibre'. The ability of hair to stretch depends upon the elongation of the α-helices of the individual keratin molecules. These can revert to their original conformations, not through the strength of their hydrogen bonds, which are broken in stretching, but because the helices are cross-linked by disulphide bonds between cysteine residues. As with all polymers, the extent of cross-linking determines the rigidity of the product. Keratins that contain little cysteine are soft, i.e. they have little cross-linking, whereas keratins with a lot of cysteine are hard. Hair is an example of the former and horn of the latter. Breaking these disulphide bonds and then re-forming them after the fibres (and consequently the keratin molecules) have been distorted is the basis of the permanent waving (or straightening!) of hair. Once the molecules have been reset they remain in that conformation – the 'wave' disappears by growth of the hair. The rather pungent odour of the liquids used in permanent waving is characteristic of disulphide bond-breaking reagents.

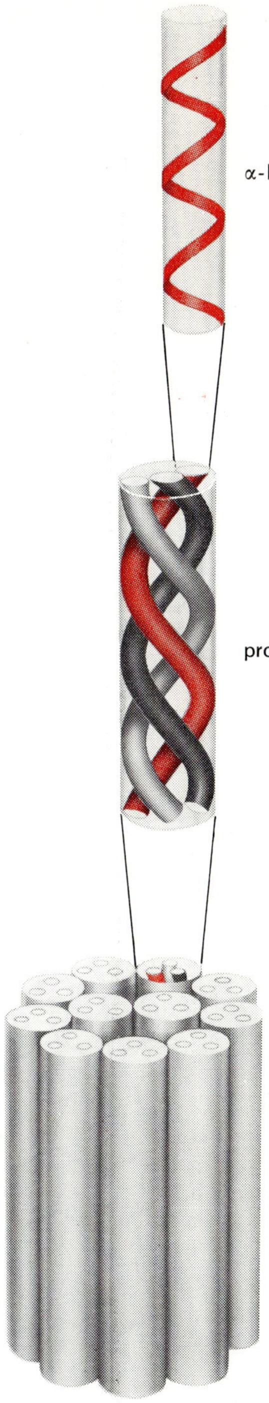

The elasticity that keratin possesses is due to the conversion of the α-helices into a stretched form, similar to that found in sheet structures. *Sheets* can be of two kinds; parallel and antiparallel, but we will deal only with the latter in our example here. In the anti-parallel sheet the side-groups of the amino acids are perpendicular to the plane of the sheet, projecting alternately above and below it. Silk, with an approximate primary sequence of

$$— (Gly–Ser–Gly–Ala–Gly–Ala)_n —$$

has all the Gly side groups one side of the sheet and all the Ser and Ala side-groups on the other. This enables the sheets to be stacked on top of one another as shown in Figure 20 (p. 24).

The polypeptide chains are almost fully extended and as a result the fibre or sheet has little capacity for stretching, although the sheets themselves, not being covalently bonded as are the chains, can bend, giving a degree of flexibility. This is a clear example of the primary structure determining the secondary structure and hence the final features of the material. Some species of silkworm produce silk fibres with properties approaching that of wool, which can stretch quite appreciably, being made of keratin. On analysis, these fibres have been found to be made from silk molecules with a quantity of amino acids other than Ser, Gly and

* *S. W. Hurry (1965), The Microstructure of Cells, John Murray.*

α-helix

protofibril

microfibril

Figure 19 *Structure of keratin fibres.*

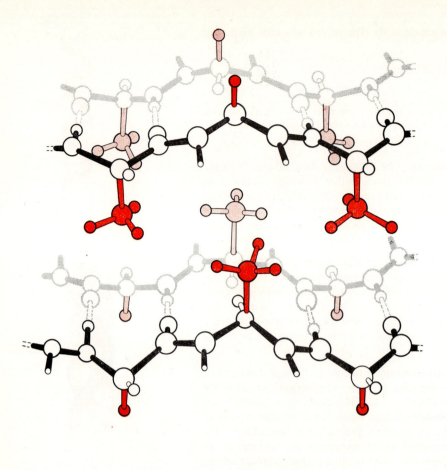

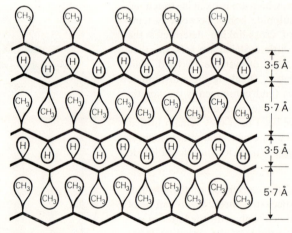

Figure 20   Structure of silk.

Ala and which are not arranged in a sheet formation. Consequently the interspersion of these extensible regions within the almost crystalline regions of sheet allow the molecules to be stretched in a similar manner to that of keratin.

Another fibrous protein, collagen, is found in structures that need to be inelastic and relatively inert. It is the basic component of tendons and of the cornea of the eye. Current knowledge of collagen structure suggests that the fundamental unit is tropocollagen, which is a fibre formed by twisting three left-handed polypeptide helices into a right-handed superhelix. Note the difference here between collagen and keratin: although both are made from L-amino acids, the α-helix in keratin is *right-handed* whereas the collagen helix is left-handed. However the collagen helix is not an α-helix; it has entirely different dimensions, being much more elongated and having 13 residues per turn as compared to 3.6 residues per turn in the α-helix. This is shown in Figure 21a and the structure of tropocollagen is schematically represented in Figure 21b.

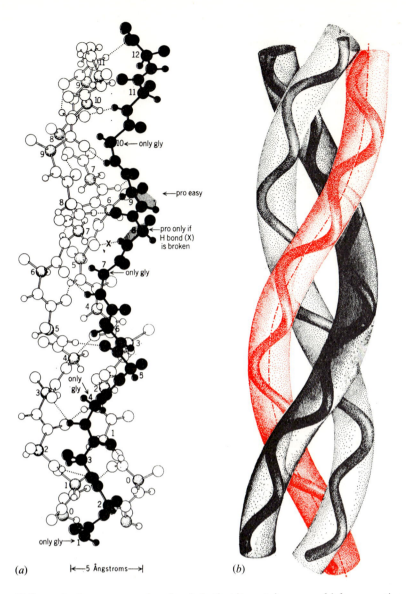

(a) |←—5 Ångstroms—→| *(b)*

Labels in figure (a): 12, 11, 10←only gly, 9, ←pro easy, ←pro only if H bond (X) is broken, ←only gly, only gly, only gly→1

Figure 21a   The basic coiled-coil structure of collagen.   Three left-hand single-chain helices wrapped around one another with a right-handed twist.

Figure 21b   Structure of tropocollagen.

Collagen is also an unusual molecule in that it contains a very high proportion of the amino acids proline and hydroxyproline (Pro and Hypro), having approximately 25 per cent of each of these and exactly 33 per cent glycine. The primary sequence has been shown to be that given in Figure 22.

Figure 22

$$— (\text{Gly–X–Pro})_n — \quad or$$

$$— (\text{Gly–X–Hypro})_n —$$

If X is Pro, then a bonding hydrogen is lost and consequently the molecule is less stable than if some other amino acid were present. We said that the presence of large amounts of Pro and Hypro makes collagen unusual compared with other proteins. Normally these amino acids are placed at the end of an α-helical section where they force it to terminate – they cannot be fitted into an α-helix. They create a bend in the polypeptide chain and are frequently found in globular proteins to turn the polypeptide chain back upon itself. If you look at their structures you will see that they are not, in fact, *amino* acids as we said above but *imino* acids, i.e. the nitrogen is bonded to only one hydrogen (imino), not two as in the amino (Fig. 9.13, p. 188, in *Cell Structure and Function*).

imino acids

1.5.3   Polysaccharides

Those polysaccharides whose primary structures have been worked out are also the few whose secondary structures are at least partly known. Two types of secondary structure have been found; linear and globular. Cellulose, the fibrous polysaccharide from plants, consists of long, fairly rigid glucose chains. In cellulose fibres these molecules lie parallel to each other and, because the glucose residues are alternatively rotated through 180° (S100[22]) can get close enough together for hydrogen bonds to be formed between their hydroxyl groups. The summation of the hydrogen bonds is so strong that the fibres are completely insoluble in water. This is obviously a necessity for molecules which form much of the structural support for cells.

> **ITQ 6** Remembering what we said earlier about the properties of fibres like wool and silk with relation to the secondary structures of their molecules, what properties would you predict for cotton fibres from the structure of cellulose?

On the other hand, glycogen, the animal glucose storage molecule, has a globular and rather unspecific structure. Undoubtedly some hydrogen bonding between the glucose residues does occur but not to a sufficient extent to generate any recognizable structural arrangements. Similarly, amylopectin, with a primary structure similar to that of glycogen, seems to have no definable secondary structure, although it is interesting to note that when amylopectin is combined with amylose (which seems to have a helical conformation) in starch granules, these exhibit a very clear regular crystalline pattern when studied by X-ray diffraction (S100[2]).

1.5.4 Some experimental approaches

So far, you have looked at all these examples of secondary structure without reference to the ways in which they were obtained. The most reliable evidence of whether a macromolecule has any secondary structure, and if so of what kind, is that derived from X-ray diffraction data. The types of secondary structure we have mentioned were deduced in the first instance from this kind of data in conjunction with information about bond angles and lengths. Prior to the use of X-ray diffraction, guesses could be made of *possible* structures using molecular models (e.g. Pauling's for the α-helix) and evidence such as ultraviolet absorption (see below), but the demonstration of the existence of regular structures needed some form of 'visualization'. X-ray diffraction provided this.

However, if precise structures are not present in every molecule of a crystal sample, then the data obtained by X-ray diffraction will be meaningless. Only with a regular array of atoms in a crystal matrix can a useful analysis be performed. It has only been possible to analyse a few molecules in this way because of the difficulties involved in preparing pure crystals of them. (This need for crystals, incidentally, raises the problem of whether the conformations of the molecules in the crystalline state are the same as those in solution.) In the absence of such data there are various other methods which can be applied to the measurement of secondary structure.

X-ray diffraction

The ordered arrangements of the atoms or groups (e.g. $C=O$ or $N-H$ in proteins) in the regular formations of secondary structures lead to phenomena such as characteristic absorptions of radiation (S24–, Unit 1). These can be exploited to give a reasonable idea of the amount and type of secondary structure present. For instance, single nucleotides absorb ultraviolet radiation strongly at a wavelength of 260 nm†, so it might be expected that a molecule like DNA, which is made from tens of thousands of nucleotides, would absorb the radiation tens of thousands of times more strongly than one nucleotide. This is not the case. The regular arrangement of the bases, which lie above one another in a plane at right-angles to the long axis of the helix ('stacking'), causes them to interact with each other in such a way as to *reduce* the absorption to much less than would result from simple summation of all their individual absorptions. The greater the degree of order in a nucleic acid, be it mRNA, tRNA or DNA, the greater is this reduction. The effect is called *hypochromism*. If the appropriate number of nucleotides is known from hydrolysis and subsequent estimation of the nucleotide products, and if the hypochromism can be determined, then the proportion of the molecule involved in base stacking and hence helical formation can be calculated.

hypochromism

26

Another result of the regular arrangements within macromolecules is their asymmetry or chirality (S100[23]).

ITQ 7   In S100, Unit 10, you met a property of chiral molecules which could be used to measure the degree of chirality. Can you remember what it is?

In the case of biological macromolecules a variation of this method is used. Instead of measuring this rotation at one wavelength, say the sodium D line, it can be measured at many wavelengths and the amount of rotation as a function of wavelength derived. This is called the *optical rotatory dispersion* (ORD). Some typical plots of ORD are shown in Figure 23. Note the large differences not only between the nucleic acids themselves but between them and the protein. The intensity of the peaks and troughs can be used as a measure of the proportion of ordered structure present and of its type, by reference to the curves produced by known structures.

ORD

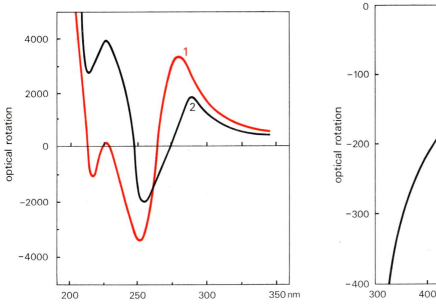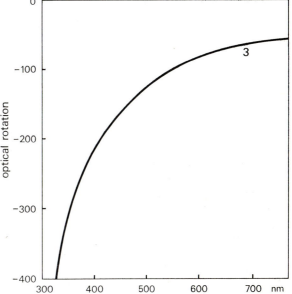

Figure 23   ORD plots for RNA, DNA and protein.

These are, of course, only the bare outlines of some of the techniques which can be used; the theoretical and practical details are far beyond the scope of this Course.

1.6   Tertiary Structure

Study Comment

Section 1.6 examines the non-regular structures, unique to each kind of macromolecule, which are called tertiary structures. It deals solely with proteins and the forces which maintain their conformations at this level, using lysozyme as an example. You should be able to remember the bond types involved and their relative importance, with some of the supporting evidence for their existence.

We have not finished with the structure of macromolecules yet. We can go further and begin to describe the molecule as a whole, to derive a much more complete although rather static picture. To do this a lot of data from X-ray diffraction is needed; not many macromolecules have been analysed in this way. At the present time the number is about 20, the majority of these being proteins.

Take a look at the three-dimensional representation of the lysozyme molecule in Figure 9–36A, p. 217, in *Cell Structure and Function*.

There is a region of β–sheet in the lower left and centre of the drawing, and areas of α–helix in the lower right and centre back. Apart from a few other marked hydrogen bonds the rest of the chain seems to be floating freely; in fact, it is not at all apparent why some parts of the chain, for instance residues 18–25, have formed in that particular way rather than any other. Why can't the molecule just bend at the lower region (called the 'cleft') to take up almost any position it likes? To find the answer to this question we will go back a little and look again at the amino-acid composition of the molecule.

In your study of primary structure you met the amino acid cysteine (Cys) and its ability to form disulphide bonds (–S–S–) with other cysteines. Lysozyme has four such bonds (not actually marked in the diagram) which all contribute to the maintenance of one unique stable conformation. (If you want to find and mark these they are between residue numbers 6–127, 30–115, 64–80 and 76–94.) You may now wonder whether this conformation in which the molecule is restrained is actually the thermodynamically most stable (S100[24]), or is it being forced into a more unstable form. Present evidence suggests the former, the experiments which were used being elegant in their simplicity.

Here is a hypothetical experiment, of the kind originally performed by Anfinsen, set up as a problem for you. Read the description of the experiment and then answer *ITQs* 8 and 9.

An enzyme has four disulphide bonds between eight cysteine residues (A to H), along a single polypeptide chain. In the original molecule the residues were connected in the following pairs:

A–B; C–D; E–F; G–H.

The enzyme was assayed for its ability to degrade a substrate, and its original specific activity was found to be 100 micromoles of substrate degraded per minute per milligram enzyme (100 μ mol S/min/mg). A sample of the enzyme was then treated with the reagents mercaptoethanol ($HS \cdot CH_2 \cdot CH_2 \cdot OH$), which is known to break disulphide bonds, and urea, which causes unfolding of the polypeptide chain by breaking its hydrogen bonds.

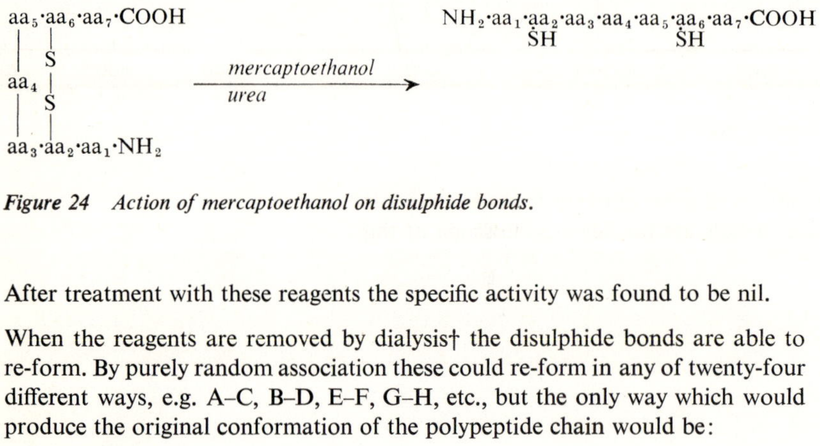

Figure 24 *Action of mercaptoethanol on disulphide bonds.*

After treatment with these reagents the specific activity was found to be nil.

When the reagents are removed by dialysis† the disulphide bonds are able to re-form. By purely random association these could re-form in any of twenty-four different ways, e.g. A–C, B–D, E–F, G–H, etc., but the only way which would produce the original conformation of the polypeptide chain would be:

> A–B
> C–D
> E–F
> G–H

The weakness of hydrogen bonds makes it unlikely that they will direct the renaturation to any great extent. They can, therefore, be ignored.

> **ITQ 8**   If this re-association *were* purely random, predict the approximate specific activity of the renatured enzyme.

> **ITQ 9**   The actual specific activity found was 94.5 μ mol S/min/mg. Can you explain the difference between the answer to *ITQ8* and this?

In the actual case of RNAase, which was the first enzyme studied in this way, random re-formation of the bonds would have reduced the activity to only 1 per cent of that of the original, whereas the actual activity was greater than 95 per cent. Admittedly this was exceptionally high, but many proteins can show 50 per cent recoveries as compared to less than 10 per cent expected; evidently in these cases the preferred conformations of the intact molecules are those in which they are enzymically active. So it would seem that although the disulphide bonds *do* maintain the structure in its active form, some other forces are also at work to make this the preferred structure.

In the molecules, there are some hydrogen bonds which are not part of any regular structures such as helices or sheets, and which could contribute to this effect, but they could not maintain the tertiary structure on their own. To find these other forces it is again necessary to look at the side-groups of the amino acids; this time, particularly those which are classified as apolar (non-polar) (S100[25]) or aromatic (S100[26]). These side-groups are unsubstituted hydrocarbons (S100[27]) and for an analogy of their reaction to an aqueous solution one could look at how other apolar hydrocarbons, for instance oil or petrol, react towards water. You know by everyday experience that petrol and water do not mix; they separate into layers or globules. They do this so that the majority of the hydrocarbon molecules do not come into contact with the water molecules and vice versa. (Petrol is therefore called a hydrophobic or 'water-avoiding' substance.) This is the most thermodynamically stable arrangement for the petrol/water system. In proteins, the most effective way for the apolar side-groups to be kept from contact with the aqueous medium is for them to be buried in the interior of the folded molecule while the polar side-groups are placed on the outside. This also avoids too much contact between the two types of side-groups (polar and apolar) which behave analogously to the water and the oil. The protein lysozyme shows this effect quite clearly and it has also been found to be true of all other proteins. However, remember that this effect is not absolute. Not *all* polar groups are on the outside nor *all* apolar groups on the inside. Indeed it is sometimes essential to the functions of an enzyme that this tendency be ignored. Apolar side-groups of amino acids can bind apolar substrates in the catalytic region ('active site') of an enzyme while the enzymic catalysis occurs. They can also markedly effect the ionization properties of polar side-groups in their immediately vicinity in the active site (see Unit 2 of this Course, Section 2.5.5).

hydrophobic bonding

When a protein is exposed to an organic solvent like acetone or alcohol the position is reversed. The apolar groups tend to go to the exterior of the molecule and the polar groups to the interior. This will obviously totally destroy the conformation, and the protein then becomes inactive or denatured (S100[28] and *The Chemistry of Life*, p. 69).

The apolar forces (or 'hydrophobic bonds' as they are often called) are believed to be the strongest single directive force involved in maintaining protein conformation. However, there are some other forces which help stabilize tertiary structures. Turn to Figure 9.30, p. 210, in *Cell Structure and Function* which is a summary of these. Note that:

(*a*) we have already mentioned hydrogen bonds, shown here between side-chains which can participate in them, and disulphide bonds;

(*b*) the figure also shows some examples of which apolar side-groups can participate in apolar forces;

(*c*) ionic bonding (which we have not mentioned) is shown between oppositely charged side-groups. Which side-groups are able to participate will depend upon the pH, since this determines which side-groups carry charges.

ionic bonding

(*d*) the iso-peptide bond (i.e. a peptide bond between *side-chains*) is an uncommon way of cross-linking a polypeptide chain. Its formation requires an enzyme which you will meet later in this Unit (in Section 1.8).

(*e*) excluding the iso-peptide bond (which we have said is rare), the only *covalent* bonding involved in maintaining tertiary structure is the disulphide linkage. In a molecule which has no Cys residues the whole of the tertiary structure is therefore maintained by non-covalent, or secondary, bonds or forces, which demonstrates the combined strength of these bonds.

As you saw in lysozyme, a protein can have regions of secondary structure interspaced with regions of so-called 'random chains'. These are not literally random, as we have shown, but are merely too complex in structure to be given a simple generic name, like 'helix'. All these irregular parts of the molecules

random chains

29

are held together by combinations of the different types of bonds we have mentioned. In the case of enzymes the whole structure is so arranged that the few groups which participate in catalytic activity are correctly positioned in the active site. In lysozyme, they are along the two sides of the cleft, which must itself be of the correct shape so that it can only catalyse reactions of its specific substrate, the polysaccharides of bacterial cell walls. The organization and action of the active site of enzymes is dealt with in Unit 2 of this Course.

It might, perhaps, seem as if some of the outer parts of the lysozyme molecules, lying far away from the active site, are somewhat redundant. However, in many cases, it is found that even these extremities do contribute to stabilizing the whole molecule. They may perform other functions as well, which are not apparent in the isolated and purified molecule. For instance, they may be capable of interacting with other molecules, maybe to bind the enzyme to a membrane or to other enzymes to form a complex. It is this idea of interaction between macromolecules which brings us to quaternary structure; the formation of a complex molecule by the combination of several different or identical subunits.

You should now attempt SAQs 3 and 5.

1.7   Quaternary Structure

Study Comment

Section 1.7 begins the study of structures with more than one macromolecule; in this case, proteins. The properties of quaternary structure are exemplified through haemoglobin which you will read about in *Cell Structure and Function*. **You should be able to define quaternary structure, give some of the advantages it has over discrete molecules, and show how closely it is determined by the primary sequences of the individual molecules.**

When a protein is composed of two or more smaller proteins it is said to be made up of subunits. (Table 9–5, p. 221, in *Cell Structure and Function* shows the numbers of subunits in several proteins. Note that there is no relation between size, i.e. MW and the number of subunits which a molecule contains.) It is useful for the sake of clarity to make an arbitrary distinction between a protein which exhibits quaternary structure and a macromolecular complex. This is most easily done from a functional rather than a structural point of view. We will define a molecule which has quaternary structure as an aggregate of subunits which are unable to perform any cellular activity (such as enzymic activity) when separated from each other, and a macromolecular complex as one composed of subunits which are able to act when unbound.

You should now read the section in Cell Structure and Function *starting at p. 223 (at the start of the paragraph 'Haemoglobin . . .') to the end of the chapter (approximately 700 words).*

Commentary on 'Cell Structure and Function' pp. 223–5

The main points to notice when reading this are:

(*a*)   the presence of 4 chains;

(*b*)   how the intereaction of these chains creates an effect not seen in a single chain and *not predictable* from it, namely the *co-operative* binding of oxygen; this increases the efficiency of haemoglobin as an oxygen-carrier;

(*c*)   the specificity of assembly of the haemoglobin molecules in the heterozygote (S100[29]) for sickle-cell anaemia.

Glossary of some terms used in the set text

α AND β CHAINS   These are the names given to the two kinds of polypeptide chains in haemoglobin. They are quite arbitrary and have no relation to primary and secondary structures.

30

ISOMORPHOUS REPLACEMENTS Addition of an easily identifiable atom, e.g. mercury, to the molecule to enable the X-ray diffraction data to be interpreted. In the process, the crystal structure must not be distorted.

MYOGLOBIN This protein is an oxygen carrier like haemoglobin but has a MW of 17 000 compared to that of 68 000 for Hb. It contains only one polypeptide chain with one atom of iron in a haem group. It is found concentrated in muscle where it acts as an oxygen store, being able to bind O_2 at much lower partial pressures (i.e. concentrations) than Hb and thus to accept O_2 from Hb in the blood. This is shown in Figure 9–40, p. 224, in *Cell Structure and Function*. It is a likely evolutionary predecessor of Hb.

SUB-UNIT INTERFACES There are four sub-unit interfaces in Hb $\alpha_1\alpha_2$; $\alpha_1\beta_1$; $\alpha_1\beta_2$; $\beta_1\beta_2$. The contacts between like chains are through polar bonding and between unlike chains through apolar bonding. The $\alpha_1\beta_1$ and $\alpha_1\beta_2$ interfaces are connected by hydrogen bonds between adjacent α-helical regions. Interface $\alpha_1\beta_1$ is a firm contact and allows no lateral movement of either chain with respect to the other, whereas the $\alpha_1\beta_2$ interface is weak and does permit these relative changes.

Haemoglobin is a fine example of how closely structure and function are knit. The alteration of one amino acid to another in the normal adult form of haemoglobin (HbA) produces the haemoglobin characteristic of sickle-cell anaemia (HbS) (S100[29]). This single change from Glu to Val at residue 6 in the β-chain creates a haemoglobin with radically different properties to that of HbA, namely that oxygenated HbS is relatively insoluble. Well over a hundred amino-acid replacements are known in human α and β chains resulting from base changes (mutations) in the DNA (S100[30]). Some of these produce large changes in the properties of the molecule and have clinical symptoms, whereas others have no deleterious physiological effects and are only detected by testing the haemoglobin of large numbers of people for abnormal amino-acid sequences ('screening'). You will meet more on this topic and see how the abnormal sequences are detected in TV programme 3 of this Course.

1.8 Higher Orders of Structure: The Cellular Components

Study Comment

Section 1.8 is in a sense the climax of the Unit. It deals with the large organelles and structures which are found in all cells. Most of this text is in the form of prescribed reading from *Cell Structure and Function*, with notes to explain which points to look out for and glossaries to deal with unfamiliar terms. You should be able to remember examples of cellular organelles, briefly what they do, and their structure. This is especially true for Section 1.8.2 on membranes. (You will meet one organelle, the mitochondrion, in Unit 3 of this Course). You should be able to remember the problems which crop up in studies of large complexes.

You have now seen how the precise structures of macromolecules enable them not only to perform specific actions, e.g. as enzymes, but also to interact with one another to form complex molecules. When this is taken a stage further one is dealing with macromolecular complexes made from molecules which are able to act independently but can perform these actions much more efficiently by co-operating with other molecules that operate in a similar 'field'. An obvious example of this would be an enzyme that catalysed a particular reaction within a metabolic pathway, say B→C in A→B→C→D. If the two enzymes which catalyse A→B and C→D are fixed in close proximity to the B→C enzyme in a complex, then the substrate A can be converted to D with a much greater efficiency than if all three enzymes were soluble, i.e. free and separate in the cytoplasm. Such an arrangement is called a '*multi-enzyme complex*' and is a common feature of metabolic pathways, especially many important ones such as fatty acid synthesis (see Unit 4 of this Course).

multi-enzyme complex

The concept of co-operation has great appeal but creates quite a few headaches for those who are interested in studying it. Suppose that the enzyme composition of a complex has been determined by breaking it down into its component macromolecules. How can the interactions of these be dealt with? If the individual enzymes are analysed there is no *a priori* reason why they should have the same conformation when free that they have when bound in the complex. The only way in which the correct biological environment can be created for all these enzymes is to reassemble the complex, which of course brings us back to square one. This situation is made worse in complexes which contain passive as well as active components, e.g. lipids, which provide structural support and a hydrophobic environment. It is very difficult to analyse the action of something which has no action at all outside the complex.

Another general problem inherent in the study of very large complexes like organelles (S100[31]) is that of outside/inside faces. This feature of organelles is very important because they often have a vectorial factor involved in their biological function, that is they will allow or enhance the passage of certain substances, like small molecules or ions, into but not out of themselves, or vice versa. Kidney mitochondria, for example, will take up sulphate ions from the cytoplasm. This problem reaches its peak with cell membranes; how can one create both an inside and an outside environment for a membrane within a test-tube?

These are some general problems to be borne in mind when you are reading about macromolecular complexes. Several are dealt with in some detail in *Cell Structure and Function*.

Now read Cell Structure and Function *from the start of Chapter 11, p. 268, to the end of the section on bacterial flagella on p. 285 (approximately 4 000 words).*

Commentary on 'Cell Structure and Function' pp. 268–85

When you are reading through this material you should notice the different ways which can be postulated for the assembly of macromolecular complexes in the cell. There is no need to remember the examples, but note how the experimental evidence leads to the conclusions about which type of assembly process is likely in each case.

Glossary of some terms used in the set text.

PYRUVATE DEHYDROGENASE COMPLEX The reaction sequence of the complex (PDC) is essentially that shown in Figure 25. By linking these three reactions

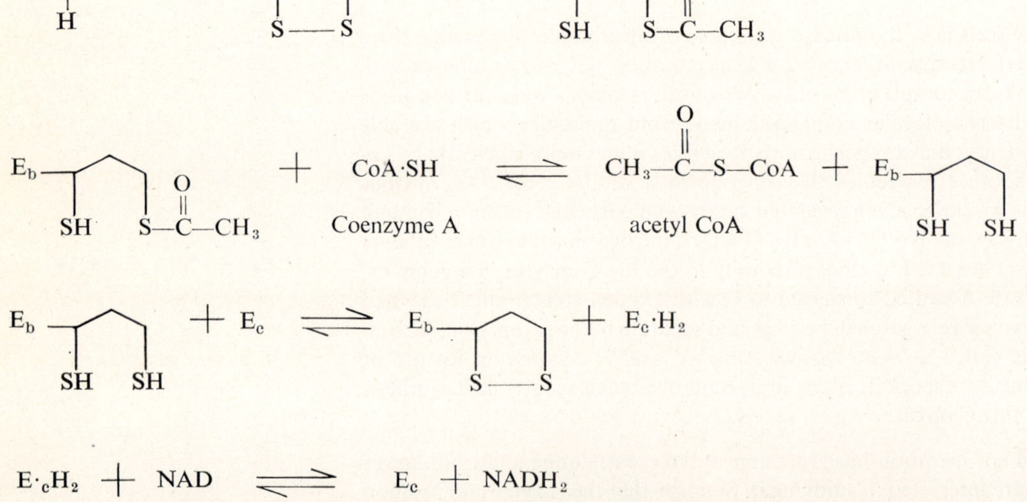

Figure 25 Reaction sequence of pyruvate decarboxylase complex (PDC).

together in one complex unit the process of converting pyruvate to acetyl CoA is dependent only upon a supply of CoA and NAD. As soon as the acetyl CoA enters the citric acid cycle, free CoA-SH is released to be recycled through the reaction sequence and so is unlikely to be in short supply. Equally NAD is not likely to limit the reaction because the complex is situated in the mitochondria where $NADH_2$ from the cytoplasm is converted into NAD through the reactions of the electron transport chain (S100[32] and Unit 3 of this Course), so normally the reaction will be very efficient.

pK   The pK of an ionizable group is the pH at which half the groups in the solution are ionized and half are not.

S VALUES   The S value (Svedberg units) of a particle is a measure of its rate of sedimentation in the ultracentrifuge under standard conditions. It depends upon the size, shape and density of the particle, a 70S particle sedimenting faster than a 35S one.

1.8.1   The membrane: a case study

You have now come across several examples of macromolecular complexes and seen how they can be broken into their components and studied. The most common general structures met in the cell are membranes, for example, the plasma or cell membranes (S100[33] and Chapter 3 in *The Microstructure of Cells*). However, very little is known about *how* they work, although a great deal is known about *what* they do. Membranes have a great diversity of function; they can control ion and substrate movements with great accuracy, repair themselves after damage and mediate the effects of hormones (S100[34]).

We will take a few of these bits of information about the roles of membranes, a few experiments and some of the theories which have been proposed and try to build a fairly up-to-date view of how membranes work.

You should now read Cell Structure and Function *from the beginning of the section entitled 'Theories of Membrane Structure', p. 469, to the end of the chapter (approximately 4 500 words).*

Commentary on 'Cell Structure and Function' pp. 469–80

The object of this reading is to show how the theories of membrane structure (unit membrane and globular organization) have been deduced from the information derived from various experiments over a number of years. The methods of fixation of tissues, e.g. freeze etching, $KMnO_4$ and the relative proportions of the various enzyme constituents need not be remembered. Note, however, the various functions which membranes have and how they can/cannot be accounted for by the two theories.

unit membrane

Glossary of some terms used in the set text

PERMEABILITY   The degree of ease with which substances can pass through a barrier, e.g. membrane.

ACTION POTENTIAL   An electrical voltage difference exists across the membrane of nerve cells (SDT 286*, Unit 1) which is produced by a difference in ionic concentrations, i.e. a higher concentration of K^+ inside the cell than outside, and higher concentrations of Na^+ and Cl^- outside than inside. This polarization can be rapidly reversed, causing a subsequent reversal of polarity across the membrane (action potential).

* *The Open University (1972) SDT 286* Biological Bases of Behaviour: *A Second Level Course, The Open University Press.*

CHEMICAL POTENTIAL GRADIENT   Transport against a chemical potential gradient involves the movement of substances from a place where they are in low concentration to another where their concentration is already high. Materials diffuse so as to create equal concentrations in adjacent places when no barrier prevents them. If they are free to diffuse, energy must be expended to maintain the gradient, and the action of doing this is called a 'chemical pump'.

OSMOSIS   This is the tendency for liquids to equalize the concentrations of substances dissolved in them on both sides of a membrane which is impermeable to the substances themselves (see Fig. 26).

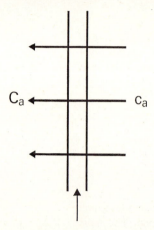

Figure 26   Osmosis across a semipermeable membrane.

If the membrane is continuous around C_a (e.g. like a cell) it will swell as water enters. If it is continuous around c_a then it will shrink. An osmometer is a device which measures the swelling and contraction of such systems.

VAN DER WAAL'S FORCES   The attractive forces between atoms which are not dependent upon charge.

PINOCYTOSIS AND PHAGOCYTOSIS   Processes whereby cells can 'eat and drink'. Pinocytosis is the process by which a cell takes in a drop of liquid in the form of a vesicle surrounded by membrane. In phagocytosis, cells 'flow around' particles and ingest them.

ZYMOGEN GRANULES   Enzymes stored in an inactive form as particles until needed.

LUMEN   A channel or passage inside the cell which is in contact with the extracellular fluid.

1.9   Some Conclusions

This Unit has of necessity been a fairly brief treatment of structure and function in macromolecules, but the main principles have all been presented. Undoubtedly the emphasis has been upon structure, partly because more is known about it and also because it underlies all attempts to explain function in molecular terms. However, few complete macromolecular structures are known in any detail, and even fewer of the larger complexes, so the field is far from cut and dried. As you read through other biological courses you will meet many other examples of macromolecular structures and you should now be better able to appreciate the depth of complexity inherent in even the simplest of them.

Self-assessment Questions

SAQ 1 (*Objective 1*) Answer *True* or *False* to each of the following:

(*a*) Proteins bear different charges because of their amino-acid side-groups.

(*b*) Nucleic acids are very easy to sequence because they have very few kinds of nucleotides.

(*c*) An endopeptidase will cleave a polypeptide chain at points within but not at the ends of it.

(*d*) All membranes have been shown to consist of lipoprotein micelles.

(*e*) The sole function of the double helix of DNA is to ensure that the molecule is inert.

(*f*) The lack of elasticity of a tendon is a direct result of the structure of its component collagen molecules.

(*g*) Variation in size of glycogen is due to the non-specificity of the enzymes which produce it.

SAQ 2 (*Objective 1*) Choose the best alternatives from the following statements:

1 The tertiary structure of a protein is maintained by:

(*a*) non-covalent bonding between acidic side-groups of amino acids;

(*b*) both covalent and non-covalent bonding between side-groups;

(*c*) hydrogen bonds between the side-groups.

2 In general, large polysaccharides are best separated from one another by:

(*a*) their difference in charge;

(*b*) their affinity for synthetic resins;

(*c*) their difference in size.

3 A macromolecule which has quaternary structure is one which:

(*a*) is composed of subunits which are individually inactive;

(*b*) has the same action as its subunits but is more efficient;

(*c*) has an action entirely different from those of its subunits.

4 X-ray diffraction analysis of DNA shows:

(*a*) what the genetic code is;

(*b*) what the structure of the molecule is;

(*c*) where the radiation emitted from the molecule comes from.

SAQ 3 (*Objective 1*) A synthetic polypeptide PBG (poly-γ-benzyl-1-glutamate) was examined by ORD in aqueous and organic solvents. In the aqueous solvent the ORD curves showed that very little α-helix was present whereas in the organic solvent they showed that the molecules were almost completely α-helical. Choose the answer below which best explains this effect.

(*a*) The α-helix is produced by apolar bonding.

(*b*) The organic solvent molecules can participate in the hydrogen bonding process and so stabilize the helix.

(*c*) The organic solvent molecules cannot compete in the hydrogen bonding process as water can.

(*d*) The effect is not explicable by the information given above.

(*e*) The water molecules interact with the peptide bonds between amino acids and break them.

SAQ 4 (*Objective 4*) A protein X was extracted from rat liver and purified until no further increase in specific activity could be obtained. It is capable of catalysing the reaction A→C which proceeds in two stages A→B, B→C. A sample of X in solution was added to a 'Sephadex' (molecular sieve) column and eluted

from it with buffer of pH 6.0. This was then repeated with another sample using a buffer of pH 8.0. The eluants were collected and assayed for protein content as they emerged from the column. From these were obtained the data shown in Figures 27 and 28.

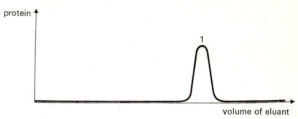
Figure 27.

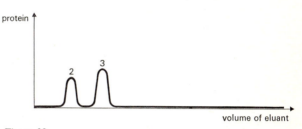
Figure 28.

The substances from peaks 2 and 3 were treated separately with urea and passed through the column again, at pH 6.0, and the protein content of the eluant recorded. This produced the data shown in Figures 29 and 30.

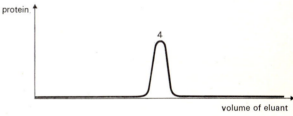
Figure 29.

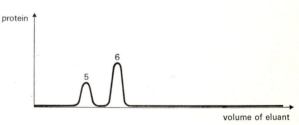
Figure 30.

When the materials from peaks 1–6 were assayed for their enzymic activity these results were obtained:

Peak 1: catalysed A→C.
Peak 2: catalysed B→C.
Peak 3: catalysed A→B.
Peak 4: catalysed B→C.
Peak 5: catalysed no reaction.
Peak 6: catalysed no reaction.

Now answer the following questions, choosing the most likely alternative.

1 The substance X is an example of:

(*a*) A single polypeptide chain protein.

(b) A multi-enzyme complex with quaternary structural subunits.

(c) A protein with quaternary structure only.

(d) None of these.

2 The substances in peaks 2 and 3 are the components of:

(a) A protein which exhibits quaternary structure.

(b) A multi-enzyme complex.

(c) A single polypeptide chain.

(d) None of these.

3 The substances in peaks 2 and 3 are most probably held together in X by:

(a) Covalent bonds of an unspecified kind.

(b) Apolar bonds.

(c) Non-covalent bonds of an unspecified kind.

(d) None of these.

4 The substance in peak 4 consists of:

(a) Subunits held together by ionic bonds.

(b) A structure of an unspecified kind.

(c) A single polypeptide chain.

(d) Subunits held together by covalent bonds.

5 The substances in peaks 5 and 6 are most likely to be held together in X by:

(a) Non-covalent bonds of some unspecified kind.

(b) Hydrogen bonds.

(c) Disulphide bonds.

(d) None of these.

6 The substances in peaks 5 and 6 are examples of:

(a) Two identical subunits behaving in different ways.

(b) Two parts of a single polypeptide chain.

(c) Subunits of a multi-enzyme complex.

(d) Two polypeptide chains which comprise a multi-enzyme complex subunit.

7 If the eluants which contained peaks 2 and 3 were mixed and eluted through the column at pH 6.0, which of the following would you expect the data to resemble most closely?

(a) Figure 28.

(b) Figure 27.

(c) Figure 29.

(d) None of these.

8 If the eluants which contained peaks 5 and 6 were mixed and eluted through the column at pH 6.0 in the presence of urea, which of the following would you expect the data to resemble most closely?

(a) Figure 29.

(b) Figure 27.

(c) Figure 30.

(d) None of these.

9 If the procedure of Question 8 were repeated at pH 8.0 in the presence of urea, which of the following would you expect the data to resemble most closely?

(a) Figure 29.

(b) Figure 27.

(c) Figure 30.

(d) Not possible to say.

SAQ 5 (*Objective 3*) A small polypeptide was isolated and purified. The sequence of the amino acids in it was determined by the use of two chemical reagents and one enzyme. You are given the modes of action of the reagents and enzyme and the data derived from the degradation process. Deduce the sequence of the amino acids.

Reagent Y reacts with free —NH_2 groups to yield a Y-amino acid. Side group —NH_2 groups *do not react*.

Reagent Z reacts with free —COOH groups to yield a Z-amino acid. Side groups —COOH groups *do not react*.

Enzyme A cleaves *all* peptide bonds on the C=O side of the amino acid PROLINE.

Total acid hydrolysis breaks *all* peptide bonds to liberate free amino acids. This process will be called TAH.

Degradation data

Treatment of the polypeptide with Y followed by TAH gave a mixture consisting of

1 part Lys-Y
1 part Glu
1 part Ala
1 part Ser
1 part Leu
2 parts Pro

Treatment of the polypeptide with Z followed by TAH gave a mixture consisting of:

1 part Lys
1 part Glu
1 part Ala
1 part Ser
1 part Leu-Z
2 parts Pro

Treatment of the polypeptide with enzyme A produced three fragments which were separated and named 1, 2 and 3 respectively.

FRAGMENT 1 TAH yielded Ala and Pro in equal proportions.
FRAGMENT 2 TAH gave Ser and Leu in equal proportions.
FRAGMENT 3 TAH gave Lys, Pro and Glu in equal proportions.

From this information you can deduce the complete sequence. ALL the information is needed.

Note The best way to approach this is to draw a scheme of all the information you have. It *should* add up to a complete answer.

SAQ 6 (*Objective 2*) The supernatant from an extract of spinach leaves with phosphate buffer† was found to exhibit an absorption of light at 560 nm proportional to its concentration. It was postulated that the substance responsible for this was iron-containing. To test this hypothesis, several spinach plants were grown from seed in a soil to which some radioactive iron (^{59}Fe) salts had been added, the plant leaves extracted as before and the substance containing ^{59}Fe separated out, using the radioactivity as a marker.

At one stage in the purification, the technique being applied was gel electrophoresis, and five different conditions were used. The results obtained from the radioactive band in the gel are shown below. From these data, answer questions (a) and (b):

(a) Which condition gives the best purification with respect to the marker being used?

(b) Is the ^{59}Fe-labelled substance responsible for the absorption at 560 nm?

condition	total protein, mg	total radio-activity, dpm	absorption/mg protein at 560 nm
1	110	38 400	2.5
2	120	29 900	1.1
3	100	35 200	1.9
4	80	35 000	1.6
5	90	33 800	0.7

(c) If the data from the stage preceding electrophoresis in the purification was as shown below, calculate how many times purer the material is at the end of the electrophoresis, using your optimum condition; also calculate the yield of ^{59}Fe-labelled material given by the electrophoresis.

total protein, mg	total radioactivity, dpm	absorption/mg protein at 560 nm
340	38 700	0.8

SAQ 7 (*Objective 5*) A macromolecular complex was extracted from liver and purified until no further increase in specific activity could be obtained. One of the measures of purity was gel electrophoresis; when the complex was run a banding pattern as in Figure 31a was obtained.

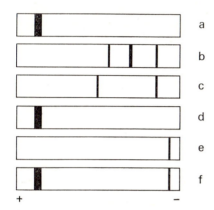

+   −

Figure 31   Stained gels after electrophoresis under different conditions.

After treatment with an anionic detergent, a sample was put through an identical electrophoresis but this time three bands were obtained (Fig. 31b). When the detergent was removed and a sample run as before, only two band were seen (Fig. 31c). However, if the 'detergent-removed' material was mixed with cell sap (i.e. cytoplasm free of large macromolecules and organelles) and repurified, it produced a band pattern on the gel as in Figure 31d.

From these data, answer the following questions:

1   Was the change which occurred upon treatment with detergent most likely to be due to:

(a) three different effects of the detergent on the complex yielding three differently modified molecules;

(b) dissociation of the macromolecule into three different subunits;

(c) dissociation of the macromolecule into two different subunits, the other band being detergent.

(d) Or is it not possible to decide between the above?

2   Is the transition between gel pattern types (b) and (c) most likely to be due to:

(a) self-assembly;
(b) aided assembly;
(c) directed assembly;
(d) none of these?

3   Is the transition between gel pattern types (c) and (d) most likely to be due to:

(a) self-assembly;
(b) aided assembly;
(c) directed assembly;
(d) none of these?

By successive purification of the cell sap it was found that the agent responsible for the transition (c) to (d) was a protein of low MW, approximately 40 000.

4   Is the transition from gel pattern type (c) to (d) most likely to be due to:

(a) self-assembly;
(b) aided assembly;
(c) directed assembly;
(d) none of these?

When the protein was subject to gel electrophoresis under the original conditions it produced a banding pattern as in Figure 31e. If the 'detergent-removed' material was mixed with this protein and the whole run on a gel, a pattern was produced as in Figure 31f.

5   Is the transition between gel type (c) and (f) most likely to be due to:

(a) self-assembly;
(b) aided assembly;
(c) directed assembly?

Answers to Pre-Unit Assessment Test

1 True. The nucleus of each atom of a radioactive isotope of an element contains one or more atoms than the stable element. The atoms disintegrate and emit particles which can be detected. (See S100, Unit 6, Section 6.3, p. 26.)

2 False. Hydrogen bonds are caused by the slight polarization of certain chemical groups which results in an attractive force (between negative and positive charges) thus:

$$= O + N\text{–}H \rightarrow\ = \overset{\delta-}{O} \overset{\delta+}{\text{—}} \overset{\delta-}{H} \text{—} N$$

(See S100, Unit 10, Section 10.5.2, p. 41).

3 False. DNA is the store for the genetic information. Messenger RNA (mRNA) is transcribed from DNA and is the molecule upon which a given protein is made. (See S100, Unit 17, Section 17.9.1, p. 32.)

4 False. A protein is made from amino acids joined together through *peptide bonds*. (See S100, Unit 13, Section 13.3.2, p. 33 and *The Chemistry of Life*, p. 57).

5 False. A molecule is hydrolysed when it is broken into two or more parts by the addition of water, e.g. the reverse of the peptide bond forming reaction (See S100, Unit 13, Section 13.3.2, p. 32, and S100, Unit 10, Section 10.4.4, p. 34.)

6 True. $pH = -\log_{10}H^+$; hence the higher the pH, the less protons in a solution, and the lower the pH, the higher the concentration of protons. (See S100, Unit 9, Section 9.21, p. 32.)

7 False. Polysaccharides are polymers of sugar molecules, e.g., glycogen is a polymer of glucose. (See S100, Unit 13, Section 13.3.1, p. 28.)

8 True. The electron transport chain enzymes are located in the mitochondrion. (See S100, Unit 15, Section 15.7, p. 45.)

9 False. Macromolecules (with the exception of DNA in undividing cells) are constantly being formed and broken down by enzymes. (See S100, Unit 15, Section 15.2, p. 8.)

10 False. Ionization can be an overall gain or loss of electrons. (S100, Unit 8, Section 8.4.6, p. 24.)

11 False. Every cell, from bacteria to mammals, is partly composed of macromolecules. (S100, Unit 14, Section 14.2, p. 13.)

Answers to In-text Questions

ITQ 1 Those proteins whose net charge was positive would move towards the negative electrode (cathode) and those whose net charge was negative towards the positive electrode (anode) (S100[6]). Assuming that frictional forces were the same on all, then their velocities would be directly proportional to their charges.

ITQ 2 Gel filtration. A gel of given pore size can be 'calibrated' with macromolecules of known molecular weights and a graph of MW versus eluant volume produced. If a macromolecule of unknown MW is passed through the gel its MW can be determined from its eluant volume by reference to the calibration curve. This only works really well for globular, i.e. spherical, molecules; long thin molecules tend to behave rather unpredictably.

ITQ 3 (*a*) Arg-Ala; Lys-Thr; Lys-Ala. (*b*) On the carboxyl side of *basic* amino acids.

ITQ 4 In a cyclic protein, i.e. one where the terminal NH_2– and –COOH have reacted to form a peptide bond. A few such cyclic polypeptides are known, e.g. tyrocidin.

ITQ 5 A molecule which has two or more identical groups attached to the central carbon atom can always be rotated to superimpose it on a mirror image of itself. See the section on chirality in S100, Unit 10. The side-group in glycine is H– so that in both the right-handed and left-handed helices the group which is closest to the C=O is H–.

ITQ 6 Cotton fibres will have little or no ability to extend. This is in fact the case. At no point in the glucose molecules or the bonds which join them can any linear extension occur without covalent bonds being broken. This is similar to the situation in silk and the converse of that in wool.

ITQ 7 The ability of chiral molecules to rotate the plane of polarized light.

ITQ 8 4.2 μ mol S/min/mg, i.e. $(\frac{100}{24})$ μ mol S/min/mg.

ITQ 9 Because the vast majority of enzyme molecules will assume the thermodynamically most favourable conformation, it would appear that, in this case, this *is* the enzymically active form. If the disulphide bonds were holding the active molecules in a form which they were, so to speak, unwilling to take up, one would expect them *not* to revert to this form after they had been released from the constraints of the disulphide bonding.

Self-assessment Answers and Comments

SAQ 1 (*a*) True. See p. 11.

(*b*) False. See p. 18.

(*c*) True. See p. 15.

(*d*) False. See p. 474 – *Cell Structure and Function.*

(*e*) False. See p. 22.

(*f*) True. See p. 24.

(*g*) False. See p. 21.

SAQ 2 1 (*b*) See p. 29.

2 (*c*) See p. 12.

3 (*a*) See p. 30.

4 (*b*). See p. 22.

SAQ 3 Correct answer: (*c*).

(*a*) is incorrect because the α-helix is maintained by hydrogen bonding between C=O and N–H groups of the peptide bonds (see p. 22). Hydrogen bonding is *polar*; that is, it is dependent upon slight charge differences between the atoms.

(*b*) is incorrect because organic solvent molecules are apolar and so cannot participate in hydrogen bonding.

(*d*) is incorrect.

(*e*) is incorrect. Water molecules cannot break or hydrolyse peptide bonds on their own. They require a catalyst such as an enzyme (see p. 15).

Water molecules can disrupt the spacings of the α-helix by being interposed between the C=O and N–H groups.

$$N\text{–}H \cdot\cdot\cdot\cdot\cdot O\text{–}H \qquad O\text{–}H \cdot\cdot\cdot\cdot\cdot O\text{=}C$$
$$\underset{H}{|} \qquad\qquad \underset{H}{|}$$
$$N\text{–}H \cdot\cdot\cdot\cdot\cdot O\text{–}H \cdot\cdot\cdot\cdot\cdot O\text{=}C$$
$$\underset{H}{|}$$

Organic solvent molecules can often enhance the *stability* of the helix by generating an apolar environment. This tends to force all the polar C=O and N–H groups into the 'core' of the helix and hence into hydrogen bonding with each other.

SAQ 4

1 Correct answer: (*b*).

The substance X catalyses two reactions, hence it is a multienzyme complex. The only difference between the two experiments was a pH change from 6.0 to 8.0 (Figs. 27 and 28 respectively). This could not have caused the cleavage of a polypeptide chain. Even if it could, a single chain would be unlikely to catalyse two consecutive reactions and even less likely to catalyse them separately when in two fragments. By definition a protein with quaternary structure has no reaction when dissociated.

2 Correct answer: (*b*).

For comments see 1 above.

3 Correct answer: (*c*).

Uncatalysed hydrolysis of peptide bonds does not occur (see 1). The only types of covalent bonds which connect separate polypeptide chains are disulphide and iso-peptide bonds, neither of which can be hydrolysed by small pH changes like this. By definition apolar bonds are not affected by pH changes, i.e., aqueous solvent effects, but non-covalent bonds like hydrogen bonds and ionic bonds could well be affected by any alterations in the charges on amino-acid side-groups.

4 Correct answer: (*b*).

The substance produced only one protein peak in the eluant from the column at pH 6.0, so no subunits have been shown to be present. However it *may* consist of subunits whose bonding is not affected by the pH changes. Hence neither conclusion can be drawn.

5 Correct answer: (*a*).

As in 3, no firm conclusion can be drawn as to which specific bond types hold these subunits together, except that they must be non-covalent. Hydrogen bonds (*b*) is too specific an answer; it may be ionic bonding.

6 Correct answer: (*d*).

The two fragments have no enzymic activity and so are not subunits of a multienzyme complex, as in (*c*).

The considerations about hydrolysis, in 1 above, exclude (*b*) as an answer. Answer (*a*) defies the basic scientific assumption that identical substances will behave identically when treated in an identical way.

The only compatible answer is (*d*), namely that there are two polypeptide chains which are joined in some way to form a subunit of a multienzyme complex, that is, the subunit has quaternary structure.

7 Correct answer: (*b*).

The alteration in pH from 6.0 to 8.0 caused X to dissociate into the two substances found in peaks 2 and 3. pH effects act upon ionizable groups, in this case amino-acid side-groups, and these changes are reversible. It is therefore reasonable to expect that when the changes in charge upon the proteins are reversed the two subunits will revert to their original state, that is will bind together. Figure 29 is derived from only one substance, that from peak 2, and that after treatment with urea. There is no reason to suppose that there will be any connection between its behaviour and that of a mixture of peaks 2 and 3.

8 Correct answer: (*c*).

There has been no change in the system. The substances in peaks 5 and 6 were dissociated at pH 6.0 in the presence of urea before, and even after mixing the eluants will still be dissociated. The peaks in Figures 27 and 29 are from different substances and/or conditions.

9 Correct answer: (*d*).

There is no data available to show how the substances in peak 3 will react to *both* urea and pH 8.0 simultaneously.

SAQ 5 The primary sequence of the polypeptide is:

NH₂·Lys·Glu·Pro·Ala·Pro·Ser·Leu·COOH

This is deduced thus:

(*a*) There are only seven amino acids in the polypeptide, Lys being N-terminal. This comes from the TAH+Y data. If there were more than seven amino-acids, e.g. fourteen, twenty-one, etc., there would still only be *one* Lys-Y, the others would be in the chain and would therefore not be labelled.

(*b*) The data from the TAH+Z process confirms the presence of only seven amino acids and shows that Leu is C-terminal.

(*c*) Enzyme A cleaves at the COOH side of all Pro residues.

Hence fragments 1 and 3 must be:

NH₂·Ala·Pro·COOH (1) and

NH₂·(Lys, Glu)·Pro·COOH (3) where the residues in parentheses are of an unknown order.

(*d*) From deduction (*b*), Leu is known to be C-terminal, hence fragment 2 is at the C-terminus with a structure of:

NH₂·Ser·Leu·COOH

(*e*) The structures of all the fragments are now known and the C– and N– terminal ones located.

NH₂·Lys·Glu·Pro·COOH +NH₂·Ala·Pro·COOH + NH₂·Ser·Leu·COOH

which gives a final structure of

NH₂·Lys·Glu·Pro·Ala·Pro·Ser·Leu·COOH

SAQ 6   Correct answers:

(*a*) Condition 4.

(*b*) No.

(*c*) 3.9 times; 23.5 per cent.

(*a*) To determine which preparation is the purest with respect to iron, one needs to know which has the highest specific activity with respect to iron, i.e. the specific radioactivity. (All calculations below are only approximate, so don't worry if your answer isn't exactly the same as ours.)

condition	specific radioactivity (*dpm/mg protein*)
1	38 400/110 = 350
2	29 900/120 = 250
3	35 200/100 = 352
4	35 000/80 = 440
5	33 800/90 = 375

(*b*) If the iron-containing substance *is* responsible for the absorption at 560 nm the specific activities with respect to both iron and absorption should vary in parallel. The values are:

condition	specific radioactivity (*dpm/mg protein*)	specific activity (*absorption/mg protein*)
1	350	2.5
2	250	1.1
3	352	1.9
4	440	1.6
5	375	0.7

These do not vary in parallel and so the two properties have not been shown to be linked.

(*c*) The purest preparation from the electrophoresis has a specific radioactivity of 440 dpm/mg protein. The purity at the end of the previous stage was 38 700/340 = 114 dpm/mg protein, and so the fold purification is 440/114 = 3.9 times.
The yield of ⁵⁹Fe labelled material is 80/340 × 100 per cent = 23.5 per cent.

SAQ 7

1   Correct answer: (*b*).

The detergent will have the same effect upon all the molecules and also, because it is an anionic detergent, it bears an overall negative charge and will not run towards the anode during electrophoresis. Therefore answer (*b*) seems likely.

2   Correct answer: (*a*).

Aided assembly requires that some other molecule than one incorporated into the complex be present. No such molecule is present in this preparation. The most positively charged molecule is part of the complex and cannot perform this role. Directed assembly requires that some of the final product be present to direct the synthesis of more of itself and, as all of the complex has been dissociated, this is not possible.

3   Correct answer: (*b*) or (*c*).

There is not enough evidence to decide between these two alternatives. The process of assembly cannot be self-assembly because it needs an external factor (the cell sap) to allow it to occur. However, until more is known about the active agent in the sap, no conclusions can be drawn as to the mechanism.

4   Correct answer: (*b*) or (*c*).

There is still not enough evidence from which to draw a firm conclusion. Aided assembly could be via an enzyme and directed assembly probably via an initiator rather than a template, the only type of the latter known being DNA/RNA. Self-assembly is not possible for the reasons mentioned in the answer to 3 above.

5   Correct answer: (*b*).

This is the most likely, although not the only answer. An initiator could be incorporated into the final product if it were a fragment of the complex. A template is ruled out because it could not be incorporated and, as the only template known is DNA/RNA, this is obviously incompatible.

References from S100*

1   Unit 14, *The Chemistry and Structure of the Cell*, p. 28.

2   Unit 28, *The Wave Nature of Light*, p. 29.

3   Unit 13, *Giant Molecules*, p. 35.

4   Unit 6, *Atoms, Elements and Isotopes: Atomic Structure*, p. 26.

5   Unit 13, p. 16.

6   Unit 9, *Ions in Solution*, p. 20

7   Unit 9, p. 32.

8   Unit 9, p. 20.

9   Unit 13, p. 32.

10   Unit 10, *Covalent Compounds*, p. 34.

11   Unit 21, *Unity and Diversity*, Appendix 3.

12   Unit 17, *The Genetic Code: Growth and Replication*, Section 17.2.

13   Unit 17, p. 31.

14   Unit 13, Section 13.3.1.

15   Unit 10, p. 41.

16   Unit 17, p. 17.

17   Unit 10, p. 17.

18   Unit 17, p. 37.

19   Unit 17, p. 32.

20   Unit 10, p. 28.

21   Unit 10, p. 26.

22   Unit 13, p. 31

23   Unit 10, p. 35.

24   Unit 11, *Chemical Reactions*, Section 1.3.

25   Unit 10, p. 9.

26   Unit 10, Appendix 5.

27   Unit 10, p. 13.

28   Unit 14, p. 15

29   Unit 19, *Evolution by Natural Selection*, p. 13.

30   Unit 19, p. 12.

31   Unit 14, p. 24.

32   Unit 15, *Cell Dynamics and the Control of Cellular Activity*, p. 45.

33   Unit 14, p. 24.

34   Unit 18, *Cells and Organisms*, p. 34.

* 1971 edition.

Glossary

AMORPHOUS   Having no shape or form (from the Greek).

ÅNGSTRÖM UNIT   The ångström unit (Å) is a small measure of length (10^{-10}m) which is extensively used in electron microscopy. Although it is not an SI unit, there is strong pressure from electron microscopists and others to retain its use.

BUFFER   A solution which maintains a specific pH regardless of how many H^+ or OH^- ions are added to it, within certain limits. Usually buffers are only effective within a few pH units of their own pH value. They are made from dilute solutions of organic or inorganic salts. For example, a phosphate buffer contains K_2HPO_4 and KH_2PO_4 and can take up or release H^+ ions thus:

$$H_2PO_4^- \rightleftharpoons H^+ + HPO_4^{2-}$$

Different salt solutions have different pH ranges over which they can operate. Below or above these ranges the ions are all in one form, e.g. all $H_2PO_4^-$ at low pH and all HPO_4^{2-} at high pH, and so respectively cannot accommodate or liberate any more protons.

COLLOID   A colloid is a suspension in a liquid of small particles of diameter between 1×10^{-9}m and 2×10^{-7}m. If the suspension has no contamination in it, it will last for several years without settling out! One of the properties of these suspensions is that of light-scattering; that is, when a beam of light is passed through a colloidal suspension, a proportion of it is deflected by the particles and hence light can be detected at right-angles to the angle of incidence (S100, Unit 28). This scattered light can be analysed for information about the size/shape of the particles.

DIALYSIS   The removal of small ions such as NH_4^+ or SO_4^{2-} from a preparation of a macromolecule. (One method of dialysis is shown in TV programme 1 and you will use it in your Home Experiment.) Often ions interfere with enzyme assays and must be removed before the specific activity of the preparation can be determined.

D- AND L- ISOMERS   This notation has *no* connection with l- and d-isomers which refer to the direction of rotation (left and right) of the plane of polarized light by optical isomers of a molecule. It is a way, still used by biochemists but now abandoned by chemists, of relating the structure of a molecule to that of one of the two stereoisomers of glyceraldehyde. Unfortunately it does not help one to work out the absolute configuration of a molecule (unless it is a very simple one) and so a new notation called R-S has been introduced. The R-S system enables one to define absolutely the configuration of *any* molecule. As this is the only place in this Course where the problem arises we have not attempted to deal with these notational systems.

NM   This SI abbreviation (nm) is used for the length of 10^{-9} m (nanometre). It is used when wavelengths of radiation are being quoted. Occasionally you may meet wavelengths measured in mμ (millimicrons), which is exactly the same length as nm, but the SI form, nm, is coming into general use. Less frequently ångström units (Å) are used (q.v.).

Acknowledgements

Grateful acknowledgement is made to the following for material used in this Unit:

Figures 19, 20, 21: Harper & Row for R.E. Dickerson and I. Geis, *The Structure and Action of Proteins*.

Contents

Table A

List of Scientific Terms, Concepts and Principles

Taken as prerequisites			Introduced in this Unit			
1 Assumed from general knowledge	2 Introduced in previous Unit, S100 or S24–	Unit No.	3 Developed in this Unit or in its set book	Page No.	4 Developed in a later Unit	Unit No.
		S100*				
vitamin	atom	5	enzyme assay	8	regulation of cellular	
muscle	litre	5	isotope dilution	11	metabolism	5, 6
temperature	Avogadro's number	6	coupled enzyme assay	12	physiological roles of	
	infra-red	6	initial rate	13	isoenzymes	5
	isotope	6	enzyme-substrate			
	mole	6	complex	13		
	molecule	6	active site	14		
	molecular weight	6	'lock-key' hypothesis	14		
	proton	6	'induced fit'	14		
	radioactivity	6	model system	17		
	spectrum	6	quasi-substrate	20		
	ultra-violet	6	pK_A	20		
	colorimeter	7	catalytic power	26		
	acid	9	proximity effect	26		
	alkali	9	acid-base catalysis	26		
	anode	9	covalent intermediate			
	buffer	9	catalysis	27		
	electrode	9	rack hypothesis	28		
	equilibrium	9	strain hypothesis	28		
	ion	9	electronic strain	28		
	pH	9	micro-environment	29		
	amino acid	10	antibody	31		
	antibiotic	10	isoenzyme	33		
	optical isomers	10	serum	34		
	activation energy	11	metabolic disease	34		
	initial rate	11	chemotherapy	35		
	thermodynamics	11	nerve gas	35		
	catalyst	12				
	protein	13	CS&F**			
	centrifuge	14	non-competitive			
	ADP	15	inhibition	243		
	ATP	15				
	activator	15				
	active centre	15				
	aerobic	15				
	anaerobic	15				
	coenzyme	15				
	cofactor	15				
	competitive inhibitor	15				
	end-product inhibition	15				
	enzyme	15				
	enzyme specificity	15				
	manometer	15				
	metabolism	15				
	NAD	15				
	optimum pH	15				
	pH-activity curve	15				

* The Open University (1971) *S100 Science: A Foundation Course*, The Open University Press.

** A. G. Loewy and P. Siekevitz (1969) *Cell Structure and Function* (2nd ed.), Holt, Rinehart and Winston.

Taken as prerequisites			Introduced in this Unit			
1 **Assumed from general knowledge**	2 **Introduced in previous Unit, S100 or S24–**	**Unit No.**	3 **Developed in this Unit or in its set book**	**Page No.**	4 **Developed in a later Unit**	**Unit No.**
	prosthetic group	15				
	reaction-time plot	15				
	substrate	15				
	gene	17				
	nitrogen fixation	20				
	evolution	21				
	X-rays	28				
	Haber process	34				
		S24–*				
	absorption of light	1, 3				
	nucleophile	3				
	strain	3				
		S2–1				
	dialysis	TV 1				
	pH meter	TV 1				
	ångström (Å)	1				
	gel	1				
	electrophoresis	1				
	peptidase	1				
	specific activity	1				
	subunit	1				

* The Open University (1972) *S24– Science: A Second Level Course, An Introduction to the Chemistry of Carbon Compounds*, The Open University Press.

Objectives

At the end of this Unit you should be able to:

1 Define, or recognize adequate definitions of, or distinguish between true and false statements concerning each of the terms, concepts and principles in Table A (*SAQ*s 1, 2, 3, 6, pp. 14, 15, 18, 30).

2 Suggest and indicate the principles of methods for following particular reactions, given details of the nature of the reactants, products and enzymes (*SAQ* 1, p. 14).

3 Evaluate and construct hypotheses to explain the reaction mechanism of specific enzymes (real or hypothetical), given suitable data (*SAQ*s 2, 4, 5, pp. 15, 29, 30).

4 Evaluate statements concerning the occurrence, mode of operation and clinical importance of isoenzymes (*SAQ* 7, p. 37).

5 Evaluate alleged social consequences of studies on enzymes (*SAQ* 8, p. 37).

Study Guide

When we wrote this Unit we assumed a knowledge, on your part, of S100. Before tackling this Unit you should therefore attempt the Pre-Unit Assessment Test given on p. 6. This test is based on S100, particularly on features relevant to this current Unit. You should check your answers against those given on p. 38 before proceeding with the text of this Unit. At the end of some sections of the text you will find self-assessment questions (*SAQ*s). We feel that you will benefit most by doing these as they arise and checking your answers against those given before continuing with the next section of the text.

Throughout the text, references to sections within this and other Units in this and other courses are given, as well as references to set books. These references are of three types:

(a) Those where you are directed to read something *at that point in the text*, bearing in mind any notes appended.

(b) Those you should read if you are unsure about the statement to which the reference is appended.

(c) Those merely given as indications of where you can find further detail *if you want or need to*. This type of reference is easily recognized by the *lack* of any instruction . . . 'to read . . . '.

You will find a number of chemical formulae and names of enzymes and substrates in this Unit. You are *not* required to remember any of them as they are only used to illustrate principles. You may, however, in some instances find that remembering some illustrative examples assists you in remembering the principles, some of which help achieve Objectives 1 to 5.

We naturally hope that you can complete this Unit and fulfil all the objectives in the time available to you. Should you run short of time, however, you may find it useful to consider the suggestions in (a), (b) and (c) below. In our opinion, the most difficult sections of this Unit are 2.4 and 2.5. These and other sections can be simplified as follows:

(a) If you find that the structured exercises in Sections 2.4.5 and 2.6 are too time-consuming, then read the commentary bit by bit, immediately after reading each set of data and interpretations.

(b) Section 2.5 can be omitted altogether, though we feel it would be a pity to do so. You can, however, complete virtually all of the objectives, except perhaps parts of 1 and 3, without this Section.

(c) Section 2.7 can be read fairly rapidly.

We naturally urge you *not* to take the above measures *unless you need to*.

You should also note that, in addition to the material contained in this text, you have to read a total of 15 pages in the set book *Cell Structure and Function**; 8 pages during Section 2.3 and 7 pages during Section 2.4 of this Unit.

* A. G. Loewy and P. Siekevitz (1969) *Cell Structure and Function* (2nd ed.), Holt, Rinehart and Winston.

5

Pre-Unit Assessment Test (for Recall of S100)

You should now attempt this test, then check your answers against those on p. 38 before proceeding with the Unit.

For Questions 1 to 8 mark the statements 'true' or 'false'.

1 A catalyst is a substance that increases the rate of a chemical reaction, but does not itself participate in the reaction.

2 Enzymes are biological catalysts found in all living cells.

3 Any enzyme can catalyse a wide variety of reactions.

4 Some enzymes are composed of protein, others are not.

5 A coenzyme is a non-protein substance required by an enzyme for catalytic activity.

6 As catalysts, enzymes can force otherwise impossible reactions to occur.

7 The active centre of an enzyme is the region that combines with the substrates.

8 The rates of enzyme-catalysed reactions are uninfluenced by pH.

9 An enzyme, E, catalyses the following reaction:

$$A+B \rightleftharpoons C+D$$

On adding A′, a substance related in structure to A, to a mixture of A, B and E, the reaction ceases. This is probably due to:

(a) A′ being a coenzyme for the reaction.

(b) A′ not being a substrate.

(c) A′ being an inhibitor.

(d) A′ being a substrate.

Choose the best alternative.

10 An enzyme, E, catalyses the following reaction of compound A:

$$A \quad \begin{array}{c} CH_3 \\ | \\ H-C-OH \\ | \\ COOH \end{array} \rightleftharpoons \begin{array}{c} CH_3 \\ | \\ C=O+2(H) \\ | \\ COOH \end{array}$$

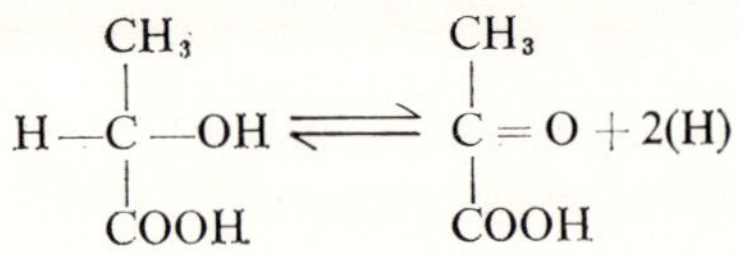

but will not act on B,

$$B \quad \begin{array}{c} CH_3 \\ | \\ CH_2 \\ | \\ H-C-OH \\ | \\ COOH \end{array}$$

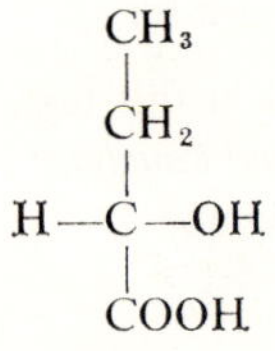

This may be because:

(a) The enzyme is highly specific for A.

(b) B is a substrate of the enzyme.

(c) B is a product of the reaction.

(d) Any one of (a), (b) or (c) is true.

(e) Some other alternative, (a), (b) and (c) each being false.

Choose the best alternative.

Now check your answers to these ten questions against those given on p. 38 before proceeding with this Unit.

2.1   Introduction: The Nature of Enzymic Catalysis

Enzymes are highly specific biological catalysts, composed of protein, which enable living cells to carry out reactions at rapid rates within a narrow range of temperature and pH.

This concise statement about what enzymes are and what they do should not be taken to indicate that everything about them is known – far from it. Relatively little is known about *how* they work. In this Unit you will learn some of what is known about enzymes and will examine some methods by which this knowledge was obtained and by which further understanding may be gained.

Consider briefly the two most fundamental characteristics peculiar to enzymes – specificity and high rates of catalysis.

As you will recall from *The Chemistry of Life** enzymes are often remarkably specific, in certain instances acting only on one of a pair of optical isomers. Thus, for example, the enzyme lactic dehydrogenase from bacteria will act on $(-)$ lactic acid to oxidize it to pyruvic acid, but has no effect on $(+)$ lactic acid:

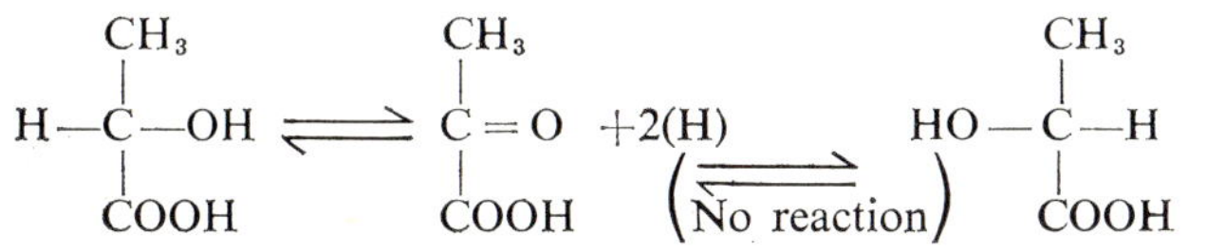

enzyme specificity

$$
\begin{array}{ccccc}
& CH_3 & & CH_3 & & & CH_3 \\
& | & & | & & & | \\
H-C-OH & \rightleftharpoons & C=O & +2(H) & & HO-C-H \\
& | & & | & & & | \\
& COOH & & COOH & \left(\overrightarrow{\text{No reaction}}\right) & COOH \\
(-)\ lactic\ acid & & pyruvic\ acid & & & (+)\ lactic\ acid
\end{array}
$$

Some lactic dehydrogenases from other organisms will only act on $(+)$ lactic acid. In addition to being selective about what substances they act on, enzymes are also very precise as to how they change such substances. Whereas one type of enzyme may convert a compound to a certain product, another type of enzyme may be capable of acting on the same compound to convert it to some other product. Both reactions must be thermodynamically possible, of course, but each enzyme, by lowering the activation energy for one reaction more than for the other, exerts its own particular directive effect. For example, glucose-6-phosphate can be converted to any one of four products; the rate at which each particular product is produced is influenced by a different enzyme:

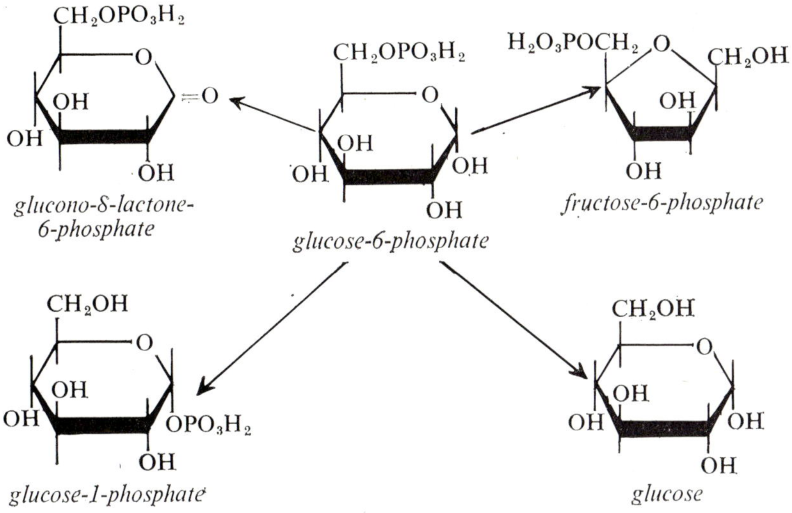

directive effects

glucono-δ-lactone-6-phosphate   *glucose-6-phosphate*   *fructose-6-phosphate*

glucose-1-phosphate   *glucose*

Thus, different enzymes can direct the same compound into different pathways of metabolism. This has important consequences for the regulation of cellular metabolism and will be considered in more detail in Unit 5 of this Course.

---

* *Steven Rose (1970)* The Chemistry of Life, *Penguin Books, Open University edition.*

Not only are enzymes highly specific catalysts but they are also very efficient ones. They frequently increase the rates of reactions by factors of 10^9 to 10^{12}. This efficiency is of fundamental importance to living organisms. By virtue of the effectiveness of enzymes, metabolic reactions occur *in vivo* at rates commensurate with life processes as we know them. Without enzymes, these same reactions would require extremes of physical conditions (incompatible with life) to produce the same rates. An example familiar to you from S100 is the fixation of nitrogen. *In vivo* – in the bacteria-containing nodules of clover, for example (S100, Units 20 and 34) – enzymes catalyse the reaction at soil temperature and pressure. In an industrial setting, the Haber process commonly uses temperatures of 400 °C, pressures of 800 atmospheres, as well as the best available inorganic catalysts.

This dual efficiency of specificity and high catalytic rate has not yet been approached by catalysts used in organic and inorganic chemistry. How then do enzymes achieve this?

This question has not yet been fully answered, despite the many years' work that has gone into the study of enzymes. In this Unit, we have attempted to present some of the established facts about enzymes and indicate how they were, and are being, established. In many instances there are few hard-and-fast answers; so we have given you some of the controversial data and ideas to allow you to judge for yourselves the scale of the problems involved.

Any study of enzyme catalysis logically starts with methods for following enzyme-catalysed reactions and these methods we discuss in the next Section (2.2). This is followed by some consideration of methods employed in identifying the active centres of enzymes (2.4), allowing us to then consider the highly controversial topic of how catalysis operates (2.5). Though not everything is understood about enzymes, some of the available knowledge has already proved useful in medicine and industry and it is with these topics that we end this Unit (2.7).

2.2 The Assay of Enzymic Reactions

Study Comment

In this Section we consider methods for following enzyme-catalysed reactions. For each method we give one or more examples. You do *not* need to remember the examples but should recall and understand the principles well enough to complete Objective 2. You should also be able to recall the names of the basic apparatus employed in such methods, as listed in Table A.

The measurement of the rate of an enzyme-catalysed reaction is termed an *enzyme assay*. The term *assay* means *measurement*. How then can one measure the amount of enzyme present in a sample by measuring the rate of reaction it catalyses? The answer to this should have been apparent in your reading of S100, Unit 15. If other factors are not limiting, the rate of the reaction is directly proportional to the amount of enzyme present – double the concentration of enzyme, double the rate, etc. It is this fundamental relationship which is the basis of all assay methods. The major practical problem, however, is that of *how these rates are measured*. In a reaction where two substances, A and B, react reversibly to give two products, C and D:

enzyme assay

$$A+B \rightleftharpoons C+D.$$

One can follow the reaction by either observing the appearance of C or D, or the disappearance of A or B. To do this, any feature which distinguishes C or D from A or B can be utilized. So in the reaction catalysed by the enzyme urease:

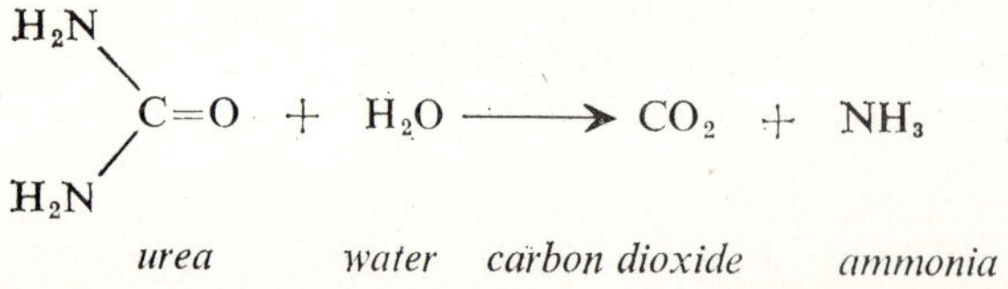

$$\underset{\text{urea}}{\underset{H_2N}{\overset{H_2N}{\diagdown}} C=O} \; + \; \underset{\text{water}}{H_2O} \; \longrightarrow \; \underset{\text{carbon dioxide}}{CO_2} \; + \; \underset{\text{ammonia}}{NH_3}$$

8

the enzyme could be assayed by the production of the pungent smell of ammonia. However, as you will recall from S100, one of the crucial requirements of scientific study is to be able to quantify one's observations. How much smell? Twice as much, half as much? Obviously the amount of smell is not measureable. Characteristics that can be accurately and reproducibly measured must be used. You can now consider some of the standard techniques that have been employed in assaying enzymes.

2.2.1 Absorption of light

In Units 1 and 3 of S24— you learnt that most organic compounds absorb light, the wavelengths at which they absorb depending on their structures, and the amounts to which they absorb depending on their concentrations and on the path-length of the light. If the substrates and products of the reaction under study absorb light at different wavelengths, then by observing the change in the intensity of absorption at one such wavelength the reaction can be followed. Where the wavelength is in the visible region, the reaction will be accompanied by a colour change and can be measured on a colorimeter, such as the one you used in S100.

Frequently enzyme-catalysed reactions do not involve coloured substrates or products. On occasions, it is possible, however, to replace the naturally occurring substrate of the enzyme by a related synthetic substance which on reaction yields a colour change. For example, phosphatases are enzymes which occur very widely in living cells and catalyse the removal of phosphate groups from a variety of molecules. Most of these reactions involve no colour change. Instead of using the natural substrates one can assay phosphatases with *p*-nitrophenyl phosphate as substrate. On reaction, this colourless compound yields *p*-nitrophenol which is intensely yellow:

p-nitrophenyl phosphate
(colourless)

p-nitrophenol
(yellow)

phosphoric acid
(colourless)

Another way of using a colorimeter to assay enzyme reactions that involve no colour change is to measure the substrates remaining or products produced at various times by some secondary reaction which gives a colour change. Thus phosphatase could also be assayed by taking samples from the reaction mixture after various times and measuring the phosphoric acid produced by adding a reagent (you need not consider its composition) which reacts with phosphoric acid to give a blue compound. You employed a similar technique in assaying amylase (S100, Unit 15); in that case, however, you sampled the reaction mixture and measured the quantity of substrate (starch) remaining by its ability to form a blue-black compound with iodine.

As instruments are available for measuring absorption in the ultra-violet and infra-red regions of the spectrum, one can readily assay reactions in which changes in absorption occur in these regions. In S100, Units 15 and 16 we described a substance called nicotinamide adenine dinucleotide (NAD) which acts as a hydrogen acceptor in certain enzyme-catalysed reactions. That is, it takes part in reactions of the type:

$$AH_2 + NAD \xrightarrow{\text{enzyme}} A + NADH_2.$$

One example is that catalysed by lactic dehydrogenase mentioned in Section 2.1:

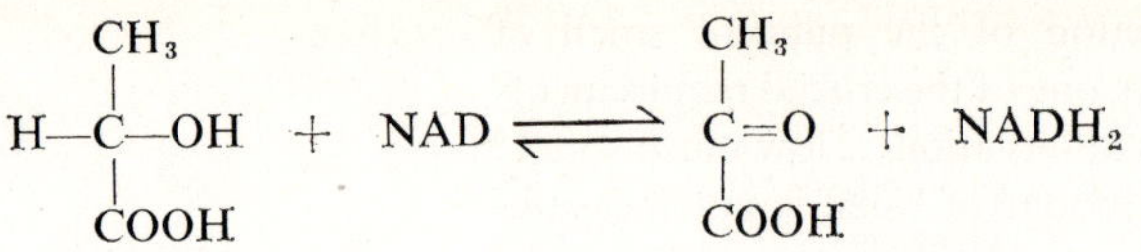

$$\underset{\substack{\text{(--) lactic acid}}}{\overset{\displaystyle CH_3}{\underset{\displaystyle COOH}{\mid}}\underset{\displaystyle }{H-C-OH}} + NAD \rightleftharpoons \underset{\substack{\text{pyruvic acid}}}{\overset{\displaystyle CH_3}{\underset{\displaystyle COOH}{\mid}}\underset{\displaystyle }{C=O}} + NADH_2$$

As $NADH_2$ absorbs ultra-violet light at 340 nm and 366 nm and NAD does not, reactions of the type shown above can be followed using an ultra-violet spectrophotometer, an instrument that measures the intensity of ultra-violet light.

2.2.2 Manometry

Other types of reaction may involve uptake or evolution of a gas and therefore, by carrying out the reaction in an apparatus capable of registering the accompanying change in volume, one can follow the reaction. You may remember our using such an apparatus – the *manometer* shown in the television programme for Unit 15 of S100.

Can you now suggest an accurate way of assaying urease (Section 2.2)?

Some reactions that do not involve gaseous uptake or production can nevertheless be followed by manometry, provided they can be coupled to a reaction leading to such a change. If reactions which produce acid are done in the presence of a bicarbonate buffer in equilibrium with an atmosphere of oxygen and carbon dioxide, the protons produced will combine with the bicarbonate anions, thus shifting the equilibrium between these and the gaseous carbon dioxide. This will result in the release of carbon dioxide gas from the buffer in proportion to the protons produced:

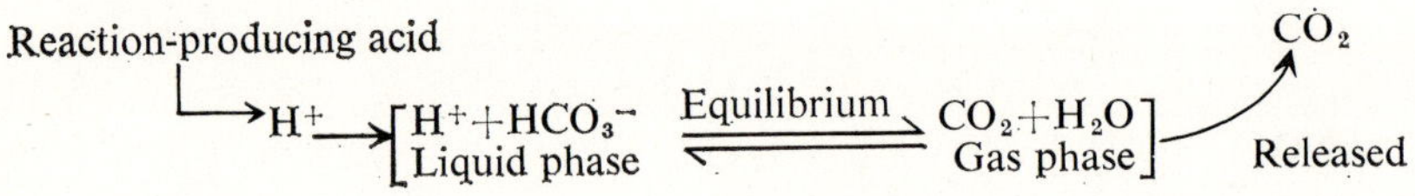

Since the reaction involves the production of two gases (ammonia and carbon dioxide) from a solid (urea) and a liquid (water) the reaction can be followed in a manometer. For each molecule of urea broken down, two molecules of gas will be evolved.

$$\text{Reaction-producing acid} \longrightarrow H^+ \longrightarrow \begin{bmatrix} H^+ + HCO_3^- \\ \text{Liquid phase} \end{bmatrix} \underset{\longleftarrow}{\overset{\text{Equilibrium}}{\longrightarrow}} \begin{bmatrix} CO_2 + H_2O \\ \text{Gas phase} \end{bmatrix} \overset{CO_2}{\nearrow} \text{Released}$$

2.2.3 pH measurement: electrode detection

Though reactions which produce acid can sometimes be followed by manometry in bicarbonate buffer, it is generally more convenient to estimate the protons produced by measuring their concentration directly. This can be done using a pH meter. You saw a pH meter being used in the first television programme of this Course.

What is the relationship between pH and proton concentration (that is, $[H^+]$)? Use this relationship to calculate $[H^+]$ for a solution having a pH of 7.4.

A pH meter essentially consists of two electrodes attached to a meter which measures potential differences. One electrode is a standard reference electrode, the other an electrode sensitive to H^+. When these electrodes are placed into the solution under study, the H^+-sensitive electrode develops a potential difference with respect to the reference electrode and this is recorded on the meter. The magnitude of this potential difference depends on the $[H^+]$. This difference is registered as $-\log[H^+]$, that is pH.

pH is defined as $-\log_{10}[H^+]$ where $[H^+]$ is the concentration of H^+ (in moles/litre). Therefore pH 7.4 corresponds to an $[H^+]$ of antilog$_{10}$ $\overline{8}.6000$, that is approximately 3.88×10^{-8} mol/l.
If you do not understand this definition and calculation look at S100, Unit 9.

If you are uncertain about the use of logarithms refer to *MAFS.**

If one is using pH change to follow an enzyme-catalysed reaction, complications can arise. If the reaction is carried out in unbuffered solution a large change in pH may alter the activity of the enzyme (as in your amylase experiment in S100).

---

* *The Open University* (1970) Mathematics for the Foundation Course in Science, *The Open University Press.*

If the solution is well buffered to avoid this, the change in [H$^+$] may be swamped and the assay rendered insensitive. A more satisfactory method involves continuous neutralization of the acid produced, by addition of alkali, thus keeping the pH approximately constant. The rate of alkali addition gives the rate of the reaction.

Electrode techniques are not limited to H$^+$ alone, as in principle it is possible to construct a wide variety of electrodes, each specifically sensitive to a particular ion. Electrodes made of special glass can measure Cl$^-$, or K$^+$, or NH$_4$$^+$, for example. So reactions involving changes in the concentration of these ions can be followed, too. Other electrodes, depending on somewhat different principles, can measure the concentration of gaseous oxygen. Still others can measure the concentration of carbon dioxide.

2.2.4   Chemical estimation and separation: isotope dilution

So far, we have illustrated two basic assay techniques: those in which the change is followed continuously, say, by a change in absorption or pH, and those where the reaction mixture is sampled and secondary colour reactions exploited, as in your amylase experiment. Sometimes the reaction cannot be followed continuously and no simple test for substrates or products exists, or the quantities involved are too small for such a test. In such circumstances, substrate or product must be isolated from each sample taken from the reaction mixture. It must be purified and the amount present measured. This can be very laborious. In addition, where the amounts present in the sample are very small, losses incurred during purification are likely to represent a large fraction of the total substrate or product present in the sample. Since one cannot be certain that the percentage loss is the same from sample to sample the accuracy of the assay will be poor. To make life easier, it is often desirable in such cases to use a technique known as *isotope dilution*.

Assume that one wanted to measure the amount of product D in a sample taken from the reaction:

$$A + B \rightleftharpoons C + D.$$

One would synthesize and purify some D in which some of the atoms were radioactive. A measured quantity of this D of known specific radioactivity (i.e. radioactivity per unit mass), is added to the sample, mixing with the non-radioactive D present in the sample; thus a dilution of the radioactive D occurs. Some D is now isolated from the sample and purified. By now measuring its specific radioactivity, it is possible to estimate the amount of (non-radioactive) D originally present in the sample. If Ds were the amount of non-radioactive D present in the sample to which was added an amount Da of radioactive D of specific radioactivity So, then the total D in the sample would then be (Ds + Da) of specific radioactivity,

$$\frac{Da.So}{Ds + Da}$$

This equals S, the specific radioactivity of the D isolated from the sample. By following this procedure the only unknown quantity is Ds. Therefore, since

$$\frac{Da.So}{Ds + Da} = S$$

by rearranging the equation one gets:

$$S(Ds + Da) = Da.So$$
$$S.Ds = Da.So - Da.S$$

And hence,

$$Ds = \frac{Da.So}{S} - \frac{Da.S}{S}$$
$$= Da(\frac{So}{S} - 1)$$

The great advantage of this method is that it yields the amount of D in the sample, irrespective of how efficient the isolation procedure is.

11

2.2.5 Coupled enzyme assays

Where no convenient techniques exist for following an enzymic reaction or chemically estimating the substrates or products, it is sometimes possible to use an additional enzyme reaction to estimate the rate of the one under study.

If the product C of reaction (1) $A + B \rightleftharpoons C + D$ is capable of reacting in another reaction (2) $C + E \rightleftharpoons F + G$ and G can be readily measured, then by coupling (1) to (2) the rate of (1) can be determined. For example, the enzyme catalysing (2) is supplied with an excess of E plus samples taken after different times from (1). Provided E is in great excess and the equilibrium for (2) is far over to the right, then all of C will be used up (S100, Unit 12), and a measure of G produced will allow the calculation of the rate of production of C in (1). In practice, one often carries out coupled assays continuously, that is by mixing the components of reaction (1) plus E and both enzymes, all together. As C is produced it is consumed at once by reaction (2), yielding G continuously.

The following system indicates how one can assay, by means of a coupled assay, the enzyme hexokinase, that is purified as shown in television programmes 1–3. Since this enzyme produces glucose-6-phosphate from glucose and ATP, one can estimate its activity by estimating the rate of production of glucose-6-phosphate. This is done by using glucose-6-phosphate as a substrate for another enzyme, glucose-6-phosphate:NADP oxidoreductase, which, as its name suggests, oxidizes glucose-6-phosphate using NADP (a substance closely related to NAD) as coenzyme. Thus for each molecule of glucose-6-phosphate produced in the hexokinase reaction, one molecule of NADP is converted to $NADPH_2$ in the second reaction. Therefore, the amount of $NADPH_2$ (which like $NADH_2$ can be estimated by its absorption at 340 nm (Section 2.2.1)), gives a measure of the extent of the hexokinase reaction:

$$\text{glucose} + \text{ATP} \xrightarrow{\text{hexokinase}} \text{glucose-6-phosphate} + \text{ADP}$$

$$\text{glucose-6-phosphate} + \text{NADP} \xrightarrow{\substack{\text{glucose-6-phosphate:} \\ \text{NADP oxidoreductase}}} \text{6-phosphogluconic acid}$$

$$+ NADPH_2$$
Measure at 340 nm

2.2.6 Choosing an assay technique

You have now read about the principles of some of the techniques used for assaying enzymes. The survey has by no means been exhaustive, but most of the techniques with wide applications have been mentioned. In practice, then, how does one decide which technique to use in any particular case? Firstly from a knowledge of the physicochemical properties of the substrates and products of the reaction one can decide, on paper at least, what technique might be suitable. There is a limit however to how far one can predict the suitability of a technique. The enzyme may be unstable at the temperature chosen, the buffer may interfere with the reaction, the products may inhibit the enzyme. As you know from the amylase experiment in S100 one must also have appropriate controls to ensure that the changes detected are actually due to enzyme catalysis. These and many other possible complications can only be determined by trial-and-error, so the 'paper chemistry' of selecting a suitable technique is only a first step towards the development of a successful enzyme assay. This success often depends on much work and on influences such as the cost of chemicals, the availability of the necessary apparatus, and the time involved. All are important; ideally the perfect enzyme assay is quick, accurate and cheap!

2.2.7 The reaction-time plot

Assuming the availability of a suitable assay for a particular enzyme, some means of expressing its activity is required. From *The Chemistry of Life*, pp. 91–2 (S100, Unit 15), you learnt that the standard procedure is to plot a graph of the extent of the reaction (products formed or substrates destroyed) against time. This would tend to give a curve like the solid line in Figure 1.

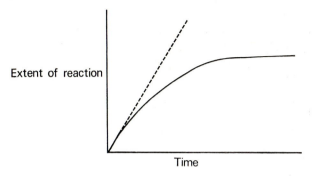

Extent of reaction

Time

Figure 1   Reaction-time plot for enzymic reaction.

Can you suggest any reasons why the reaction appears to tail-off?

Because of this tailing-off, what one considers as being the rate of reaction depends on what part of the curve one is looking at. Normally the tailing-off is less dramatic than that shown in Figure 1, which has been exaggerated for the sake of clarity.

There are many possible reasons, including the enzyme becoming inactivated during the reaction or its action is inhibited by the products of the reaction. The substrates may be unstable; or may be used up, causing the rate of the reaction to drop. The reaction could stop because the equilibrium point of the reaction is reached (S100, Units 11 and 12).

However as the reaction is less likely to be affected in its early stages by any of the factors listed above, the rate of the reaction is defined as being equal to the *initial rate* (or initial velocity). This is obtained by constructing the tangent to the curve at the time zero point (broken line, Fig. 1) and calculating the rate based on this tangent (S100, Unit 11). The rate (extent of reaction per unit time) can be conveniently expressed as x units per minute, where x is the quantity of products or a suitable measure of these (e.g. absorption units, cm^3 gas, etc.). The activity shown by a particular preparation of enzyme can be expressed in various ways: for instance as initial rate per cm^3 of enzyme solution or more meaningfully as initial rate per g of protein (the so-called *specific activity;* Unit 1 of this Course). It is the *specific activity* of yeast hexokinase that we refer to in TV programmes 1 and 2.

initial rate

specific activity

2.3   The Mechanism of Action of Enzymes

A fundamental concept about catalysts, enzymes included, is that they cannot act at a distance. To influence the rate of a reaction the catalyst must itself be a participant in that reaction. Unlike the other reactants, at the end of the reaction the catalyst is regenerated. It therefore seems reasonable to propose that the first step in the conversion of a substrate, S, to products, P, catalysed by the enzyme, E, is the formation of a complex between E and S:

$$E+S \rightleftharpoons ES$$

Some of the evidence to support this idea is discussed in *Cell Structure and Function*, from the bottom of p. 237 ('The enzyme substrate complex') to the middle of p. 244 (that is up to, but not including, 'Specificity of enzyme action'). You should now read this, but note the following:

1   Do not try to remember the Michaelis-Menten equation. Observe how it supports the existence of ES.

2   Similarly, there is no need to remember the Lineweaver-Burk equation. It is important to note that it allows one to identify whether a substance is a competitive inhibitor. If it is, this suggests that it acts somewhat like the substrate or at the same site on the enzyme as the substrate.

13

3   Note that completely non-competitive inhibition mentioned on p. 244 is rarely observed.

4   Do not consider 'allosteric enzyme inhibition', mentioned on p. 244. We shall return to this in Unit 5.

As enzymes are normally very much larger than their substrates, it is reasonable to suppose that only a relatively small portion of an enzyme can be in contact with the substrate in the ES complex. This small portion of the enzyme is termed the *active centre* or *active site*. As we discuss later in this Unit, small chemical alterations occurring at the active site can lead to a complete loss of catalytic activity. This suggests that the active site is a finely shaped region into which the substrates can fit and be reacted upon:

active site

$$E + S \rightleftharpoons ES \rightarrow E + P.$$

The fineness of shaping of the active site is supported by the high specificity of enzymes for their substrates. This rigid 'lock and key' relationship of enzyme to substrate was developed by the great German chemist Emil Fischer in 1894. While it explains enzyme specificity and competitive inhibition there are certain facts that it does not readily explain.

Presumably very small substances could always enter a 'lock-type' site. So one might expect water to be extraordinarily reactive in many enzymic reactions, owing to its small size. However, this is far from the case in practice, suggesting that active sites must really be more complex than 'locks'. To explain this and certain other observations, Koshland has modified the 'lock and key' hypothesis. His view is that not only does a substrate have to bind in the active site but it must then cause a specific change in the shape of the site so, as it were, to 'activate' the enzyme and make catalysis possible. So competitive inhibitors or very small molecules, though able to enter the active site, cannot cause the change in shape in this site required for catalysis. As will be seen later (Section 2.4.4), some evidence now exists to support this 'induced fit' hypothesis.

'lock and key'

'induced fit'

Self-assessment Questions for Sections 2.2 and 2.3

SAQ1 (*Objectives 1 and 2*)

In Table 1 you are given some details of the physicochemical nature of a series of compounds, A, B . . . In Table 2 you are given a number of enzyme-catalysed reactions involving some of these compounds. For each reaction in Table 2, you should select from List A the principles on which to base an assay of the reaction. From List B you should select suitable apparatus for each assay chosen.

Where a coupled assay is to be used, indicate which one and choose from Lists A and B the appropriate items. More than one answer can be found for several of the reactions – give alternatives.

List A	List B
(i) Coupled assay.	(1) Manometer.
(ii) Colour reaction of substrate.	(2) pH meter.
(iii) Colour reaction of product.	(3) Colorimeter.
(iv) Colour of substrate.	(4) UV spectrophotometer.
(v) Colour of product.	(5) Centrifuge.
(vi) Ultra-violet (UV) absorption of substrate.	(6) Dialysis tube.
(vii) UV absorption of product.	(7) Automatic acid neutralizer (that is,
(viii) Measurement of gas evolved.	an apparatus that neutralizes acid
(ix) Measurement of gas taken up.	by automatic addition of alkali at
(x) Measurement of H^+.	a rate which is measured.)

14

Table 1

Compound	Solid	Gas	Solution in water coloured (Yellow)	Solution absorbs UV light at 340 nm	Reacts with dye to give blue colour
A	+				
B	+			+	
C		+			
D	+				
E	+				
F	+				
G	+			+	
L	+				+
N	+		+		
P	+				
Q		+			
R	+				
S	+				
Z	+				+

+ Indicates that a compound has the property in that column.

Note All the solids are soluble in water. The gases are not. H^+ is a proton.

Table 2

	Reaction
(a)	$B \rightleftharpoons P$
(b)	$B + F \rightleftharpoons G + L$
(c)	$N \rightleftharpoons L$
(d)	$B + A \rightleftharpoons G + R + H^+$
(e)	$B + S \rightleftharpoons L + R + H^+$, where enzyme is inactive at acid pH
(f)	$G + F \rightleftharpoons A + C$
(g)	$L + G \rightleftharpoons Z + F$
(h)	$C + S \rightleftharpoons Q + P$

Note All reactions are enzyme-catalysed and reversible, but substrates are regarded as those substances on the left-hand side, products on the right, of the equations.

SAQ 2 (*Objectives 1 and 3*).

On addition of increasing amounts of substrate, the rates of all enzyme-catalysed reactions increase until they reach a maximum value.

(*a*) This is because of the Michaelis-Menten equation.

(*b*) This is because more and more enzyme molecules are operating until all are fully occupied with catalysis.

(*c*) This is because at high concentrations the substrate is an inhibitor of the enzyme.

(*d*) This is because all the substrate is used up.

Choose the best alternatives (more than one may be possible). Then check your answers to these two questions against those on p. 39.

2.4 The Study of Enzyme Mechanisms

Study Comment

In this Section, after considering what is known about the roles of activators, coenzymes, prosthetic groups and vitamins, we go on to discuss what this knowledge and certain techniques can tell us about the active sites of enzymes. You should remember the roles of the substances named above [Table A] and the principles of the techniques for studying active sites, in order to fulfil Objectives 1 and 3. There is no need to remember the specific examples or formulae given.

In attempting to understand how any enzyme works, there are essentially three questions to be answered:

1 What is the nature of the chemical changes to the substrates?

2 What is the chemical identity of the active site?

3 How do the chemical groups at the active site effect catalysis?

To answer (1) the techniques used in investigating the mechanisms of organic reactions are used (S24—) and we shall not deal with them here. If you have not studied S24—, you will not find any serious difficulty presented by not knowing the techniques used to answer Question (1). In this Section (2.4 to 2.4.5) we consider some of the methods used to answer (2) and then go on in the next Section (2.5 to 2.5.5) to describe what are as yet only guesses at the answer to (3).

2.4.1 Enzyme cofactors

All enzymes are proteins. Therefore the active site must be shaped by a three-dimensional array of the side-groups of the amino acids in the site. However, some enzymes also contain non-protein parts. As you learnt in S100, Unit 15, such substances are called activators or coenzymes or prosthetic groups. As a whole, they can be called *cofactors*, that is non-protein substances other than the substrates which are vital to the activity of certain enzymes. Cofactors that you have already come across are NAD, and some metal ions such as Mg^{2+}.

cofactors

The distinction between coenzymes and prosthetic groups is very arbitrary. Various biochemists use the terms interchangeably and the whole distinction is very blurred. Therefore we shall call all cofactors, *cofactors*, and not attempt to draw any artificial divisions. The role of a cofactor is to assist in the catalytic activity of an enzyme. Therefore cofactors undergo alteration during the enzymic reaction. Sometimes they are regenerated at the end of the reaction, or they are sometimes altered (e.g. $NAD \rightarrow NADH_2$). As they often actually react with the substrates, cofactors are located at the active sites of enzymes. The relationship between active site, cofactor and substrate is schematically shown in Figure 2.

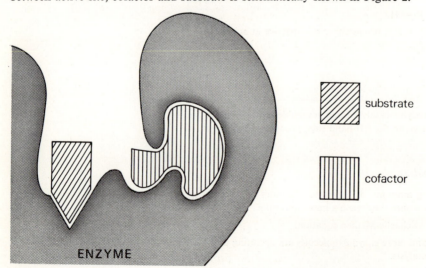

substrate

cofactor

Figure 2 Substrate and cofactor at the active site.

As cofactors can sometimes be separated from the protein portion of the enzyme and as they are relatively simple molecules, it is possible to study how they are affected by the reaction and hence learn more about the active sites of the enzymes that require the cofactors.

For example, enzymes which catalyse the interconversion of $(+)$ and $(-)$ amino acids (racemases),

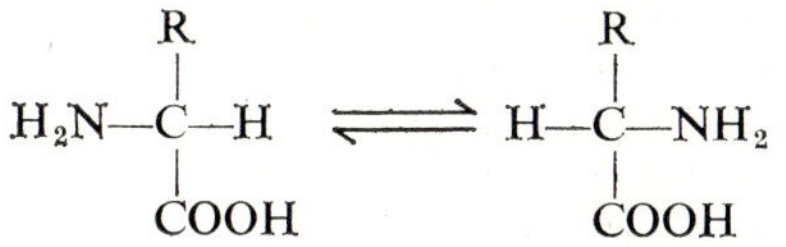

$$\underset{(+)\ amino\ acid}{H_2N-\underset{\underset{COOH}{|}}{\overset{\overset{R}{|}}{C}}-H} \ \rightleftharpoons \ \underset{(-)\ amino\ acid}{H-\underset{\underset{COOH}{|}}{\overset{\overset{R}{|}}{C}}-NH_2}$$

the transfer of amino groups between amino acids and oxoacids, such as glutamic and oxaloacetic acids (transaminase),

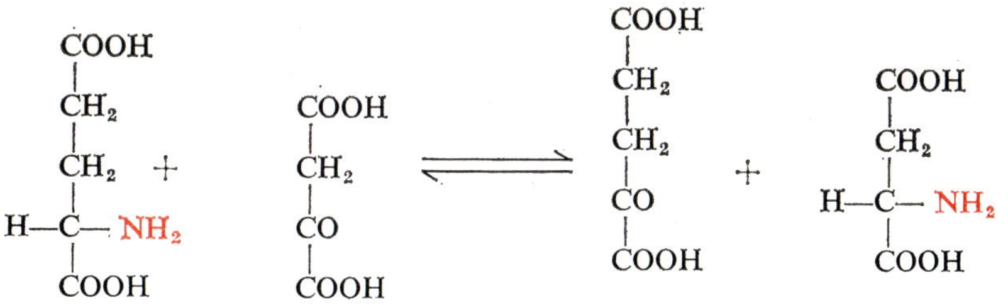

$$\underset{glutamic\ acid}{\overset{COOH}{\underset{|}{\overset{|}{CH_2}}\atop\overset{|}{CH_2}\atop H-\overset{|}{C}-NH_2\atop\underset{|}{COOH}}} + \underset{oxaloacetic\ acid}{\overset{COOH}{\underset{|}{\overset{|}{CH_2}}\atop\overset{|}{CO}\atop\underset{|}{COOH}}} \rightleftharpoons \underset{\alpha\text{-}oxoglutaric\ acid}{\overset{COOH}{\underset{|}{\overset{|}{CH_2}}\atop\overset{|}{CH_2}\atop\overset{|}{CO}\atop\underset{|}{COOH}}} + \underset{aspartic\ acid}{\overset{COOH}{\underset{|}{\overset{|}{CH_2}}\atop H-\overset{|}{C}-NH_2\atop\underset{|}{COOH}}}$$

and the decarboxylation of amino acids (decarboxylases),

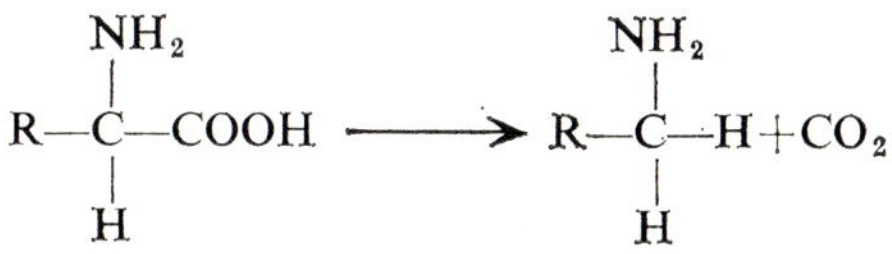

$$R-\underset{\underset{H}{|}}{\overset{\overset{NH_2}{|}}{C}}-COOH \ \longrightarrow \ R-\underset{\underset{H}{|}}{\overset{\overset{NH_2}{|}}{C}}-H + CO_2$$

all need pyridoxal phosphate as a cofactor.

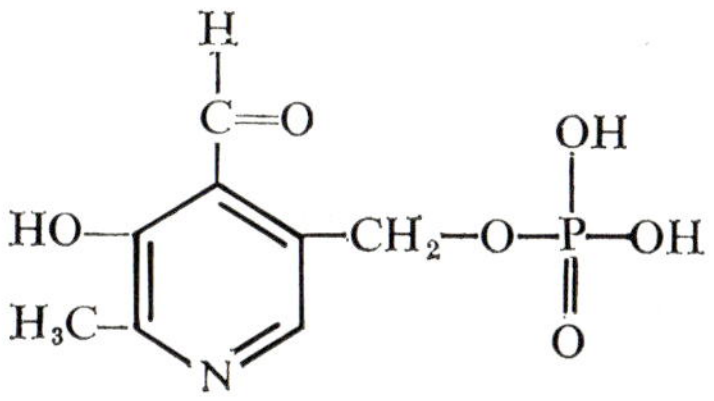

pyridoxal phosphate

It is possible to catalyse some of the reactions shown above by adding pyridoxal phosphate plus a metal ion (without enzyme) to the substrates shown. This argues strongly for pyridoxal phosphate being at the active site in the enzyme-catalysed reactions. In the simpler non-enzymic system, it is easier to study the exact nature of the complexes formed between the pyridoxal phosphate and the substrates, and thus by inference learn more about the enzymic reactions. Such systems are termed *model* systems, as it is supposed that they can be regarded as models for explaining the more complicated enzyme systems. With pyridoxal phosphate and a metal ion model system, the rate of catalysis achieved is much less than that achieved by the enzymic systems which, of course, involve the same two cofactors. This means that influences other than pyridoxal phosphate and the metal ion are important in the enzymic reactions, a concept which is reinforced by the fact that these enzymes also exert a directive effect though they all use the same type of cofactors, e.g. the enzyme catalysing the interconversion of $(+)$ and $(-)$ amino acids will catalyse *this* reaction rather than decarboxylate. So although it is relatively easy to learn much about the chemical interconversions in enzymes that use cofactors, this in itself barely paves the way towards an understanding of how these enzymes operate.

A good deal of what is now known about the role of cofactors ties up with studies from a quite different field – *nutrition*. Anyone who has read the labels on cans of fruit juice or watched television commercials is familiar with *vitamins*. These are organic substances present in foodstuffs that are needed by the body in small amounts in addition to an adequate supply of carbohydrate, protein, fat and mineral salts. Deficiency of any particular vitamin can lead to illness. Deficiency diseases are as old, or older, than man himself. Examination of skeletons of prehistoric man reveals signs of scurvy, a disease caused by a deficiency of vitamin C which leads to a bleeding of gums and membranes. The recognition of such diseases is also ancient – the curing of blind Tobias by means

vitamins

17

of fish bile, as mentioned in the Bible, suggests a knowledge of night-blindness (caused by a deficiency of vitamin A) and some therapy. By the mid-sixteenth century, oranges and lemon juice were known as cures for scurvy and in 1720 the Austrian physician Kramer wrote . . . 'if you have oranges, lemons, citrons, or their pulp and juice preserved with whey in cask, so that you can make a lemonade, or rather give to the quantity of 3 to 4 ounces of their juice in whey, you will, without other assistance, cure this dreadful evil'. By 1804 issue of lemon juice was compulsory in the British Navy, thus making scurvy a comparatively rare disease among the sailors. (Issue of lemons or limes probably gave rise to the U.S. nickname of 'limeys' for British sailors, later extended to soldiers and then to the British generally.)

In the Japanese Navy, another disease, beriberi (a form of paralysis, due to degeneration of nerves) was very common until, in 1885, Takaki, convinced that it arose from the eating of polished rice (rice without husks), as the chief article of food, managed to have some of the rice replaced by barley. The disease virtually vanished in the Navy. Presumably something in the rice husks was needed in the diet. A great step forward was taken in 1897 when Eijkmann reported that by feeding polished rice to hens he had produced experimentally a condition in these birds resembling beriberi. By 1912, Funk believed he had isolated the anti-beriberi factor which he termed the 'beriberi vitamine' (as he believed it to be *vital* and an *amine*), thus giving us the word *vitamin*. It is interesting to note that many vitamins are not in fact amines, despite Funk's terminology.

Similar studies on the feeding of controlled diets to animals, notably by Hopkins in England and Mendel and Osborne in the United States, made the existence of other vitamins clear. Many such vitamins, first known by letters and numbers, have now been isolated, purified and their chemical compositions determined. What then do vitamins do?

At first sight this is a puzzling question. Vitamins include a large number of chemically unrelated substances and what is a vitamin in one animal is not necessarily a vitamin in another. On closer examination of the chemical nature of individual vitamins it is evident that they are often chemically related to co-factors. Thus *niacin* (a vitamin which prevents pellagra, a disease leading to dry skin and nervous disturbances) is found to be *nicotinic acid*, which is closely related to NAD and NADP; the anti-beriberi substance, vitamin B_1 is *thiamine*, a substance related to a cofactor known as TPP (thiamine pyrophosphate). We are therefore led to the conclusion that some animals, ourselves included, cannot synthesize all the cofactors they require from the simple sugars, amino acids and fats that occur in food. In addition, they need from their food some more complex building blocks – vitamins. However, there is as yet insufficient evidence to support the idea that all the symptoms characteristic of a vitamin deficiency disease arise from a lack of cofactors resulting from the lack of the vitamin in question.

It is clear that diseases arising from vitamin deficiency can be cured easily if treated early enough, and can be prevented easily, that is, easily from the medical standpoint; there are many other complicating issues. A recent study has implied that even in Britain, a scientifically advanced country, some people receive less than the recommended minimum level of vitamin C. In many parts of the world people are not merely suffering from vitamin deficiency but from a lack of carbohydrate, protein and fat, as well. Scientific knowhow alone cannot eradicate problems such as these. Scientific advance needs accompanying advances in wealth, education and welfare services, before it can be used to the greatest possible benefit.

Self-assessment Question for Section 2.4.1

SAQ 3 (*Objective 1*)

Classify each of the following statements as true or false:

1 Vitamins are cofactors, because they are tightly bound to enzymes.

2 All living organisms need the same vitamins.

3 Metal ions are cofactors for some enzymes.

4 Vitamins are never metabolized by animals.

5 NAD is chemically altered by some enzymes.

Now check your answers against those on p. 39.

2.4.2 Chemical modification of active sites

The active site of an enzyme consists of a cluster of amino-acid side-groups in the protein, orientated to form a specific three-dimensional region into which substrates and cofactors (if needed) can fit. The amino-acid side-groups in the active site can be operationally subdivided into three classes, as shown diagrammatically in Figure 3.

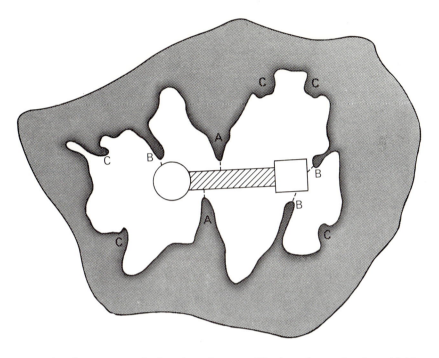

Figure 3   The active site of a hypothetical enzyme. The three classes of amino-acid side-groups are:

1   Those directly involved in the bond-breaking and bond-making steps in the catalysis (A). (The substrate is shown as a circle attached to a square. The bond to be broken is shown schematically as the cross-hatched bar.)

2   Those not involved in catalysis as such, but in contact with the substrates or cofactors, thus providing specificity (B).

3   Those present in the active site region, but not involved in catalysis at all (C).

One of the first steps towards the understanding of how any enzyme works is to identify the amino-acid side-groups in the active site and their orientations to each other and to the substrates and cofactors.

One approach is that of chemical modification: a chemical 'label' is attached to the active site, the protein is hydrolysed, and the amino acids carrying the 'label' identified. But how can one be sure that the 'label' was attached to the active site and not to some other part of the protein? Ideally a substrate of the enzyme would be used as the 'label', but, in practice, this has rarely been possible because the association between substrate and active site is too brief to survive the lengthy procedure involved in isolating a 'labelled' amino acid. On occasions, there are ways of circumventing this problem.

chemical modification

For example, incubating the enzyme aldolase with a substrate, *dihydroxyacetone phosphate*, which contains some radioactive carbon atoms, in the presence of a strong reducing agent, sodium borohydride, leads to the stable incorporation of radioactivity into the protein. Presumably reduction, by sodium borohydride, of the enzyme-substrate complex has altered this complex so as to strongly bond some part of the active site to some portion of the radioactive substrate. By now hydrolysing the protein, the amino-acid side-group responsible for the binding of the substrate could be identified. Hydrolysis reveals that a derivative of the amino acid lysine is radioactively labelled. As lysine contains an amino group in its side-group and dihydroxyacetone phosphate has a carbonyl group, it seems that the initial binding of the substrate to the lysine side-group in the active site involved a nucleophilic attack by the amino group of lysine on to the carbonyl carbon, leading to the formation of an unstable Schiff's base (details of

Schiff's base formation are given in *Basic Organic Chemistry**, p. 73). This can be represented as follows:

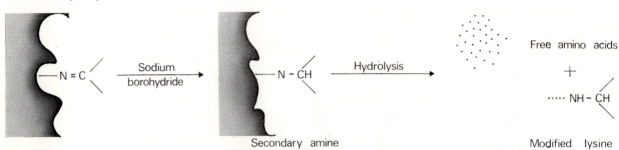

Schiff's base

The unstable Schiff's base is then reduced to a stable secondary amine, which is resistant to hydrolysis:

Secondary amine

Free amino acids

+

Modified lysine

Thus lysine is identified as an amino acid in the active site of aldolase which is responsible, at least in part, for binding substrate.

A somewhat analogous approach involves the use of so-called *quasi-substrates*. In this, one chooses a substance which reacts with the enzyme as does the substrate, thus forming a complex like ES. This complex, unlike the normal ES, is stable and does not break down to give enzyme plus products. The protein can therefore be hydrolysed and the quasi-substrate found attached to a particular amino acid. The following flow diagram illustrates the steps in the procedure:

quasi-substrates

ENZYME + QUASI-SUBSTRATE

↓

Stable complex of Enzyme-quasi-substrate

↓ Hydrolyse protein

Free amino acids + quasi-substrate-amino acid complex

↑

Identified

2.4.3   pH optima

From your amylase experiment (S100, Unit 15), you probably discovered that pH can affect enzyme activity. That is, for each enzyme there is a particular pH at which its activity is maximal, the *optimum pH*. The existence of an optimum pH is readily explicable when one considers enzymes as proteins (Unit 1). Proteins contain a large number of ionizable groups, derived from the side-groups of some amino acids (*Cell Structure and Function*, p. 188). Since the ionization of a group depends on its pK_A**, at any particular pH some groups will be ionized and some will not. Altering the pH of the solvent therefore alters the ionization of these groups. This is, in effect, a simple form of reversible chemical modification, If one or more groups at the active site need to be ionized for catalysis to occur then the pH of the system will affect the activity of the enzyme. Thus, if the

pK_A

* *J. M. Tedder and A. Nechvatal (1966) Basic Organic Chemistry, Volume 1, John Wiley.*
** *The pK_A of an acidic group may be defined as the pH at which it is 50% ionised. A fuller treatment of pK_A is given in Appendix 1 (White).*

20

relationship between pH and activity (the pH–activity curve, *The Chemistry of Life*, p. 93), is known for an enzyme and the pK_A values for the ionizable groups of the twenty types of amino acid are also known, it seems a simple matter to identify the amino acids which are involved in catalysis. Simple, it may seem, but there are possible complications, such as:

1   Several different ionizable groups with somewhat similar pK_A values might be ionized at the optimum pH.

2   The pH might affect the state of the substrates and cofactors also.

3   The pH may affect the overall shape of the protein, affecting the active site as a secondary consequence.

4   The pK_A of a group on a free amino acid may differ from that of the same group when in a protein, because of the influence of neighbouring groups.

This last point is very significant and will be elaborated later.

Much caution must therefore be exercised in the interpretation of pH-activity data.

2.4.4   X-ray crystallography

If one wishes to know the amino-acid side-groups present and the shape of the active site, a logical approach is to 'take a picture'. As you have learnt already (Unit 1 of this course) 'pictures' can be taken of proteins and three-dimensional models built from them by applying the technique of X-ray crystallography (S24—, Unit 2, and the television programme for S100, Unit 28). If this is done for an enzyme the problem of identifying the active site in the 'picture' still remains. This has recently been solved for a few enzymes by crystallizing them with their substrates (or related substances) bound.

You should now read a simple account of the elegant studies on the enzyme lysozyme, in *Cell Structure and Function*, pp. 254–8 and pp. 216–17, but note the following:

1   The figure referred to as Figure 9–37 on p. 254 is, in fact, 9–36A (shown on pp. 216–17).

2   Do not try to remember any of the details of the structures shown. Just see how X-ray analysis allows one to locate the substrate (shown in Fig. 10–19 (p. 256) as heavy black lines) in its active site.

3   We, unlike Loewy and Siekevitz, do not urge you strongly to read D. C. Phillip's article in *Scientific American*. If you wish to by all means do so.

These X-ray techniques have in the last five years or so yielded a great deal of information particularly where 'pictures' of the enzyme with and without substrate have been obtained. Such a study on the enzyme carboxypeptidase A has shown that on binding the substrate a change in shape occurs in the active site of the enzyme. This and other data lend some support to Koshland's 'induced fit' hypothesis (Section 2.3). However, before one could assume that his hypothesis was generally true, it would be necessary to have many more examples of changes in the shape of enzymes on binding substrate and show that such changes are necessary for catalysis.

X-ray analysis of enzymes is difficult and time-consuming. Confirmation is needed that what is observed in a protein crystal is also true for enzymes in solution. It would be invaluable to be able to take 'pictures' of the enzyme during catalysis – X-ray analysis only gives us 'stills'. However, these 'stills', coupled with other evidence, can enable one to build up an idea of the dynamic events that occur during enzyme catalysis.

2.4.5   Interpretation of data on enzyme mechanisms

By now you have probably realized that no one technique can give total information on how any enzyme works. A combination of the different techniques is therefore necessary. Even then, each set of evidence must be scrutinized very critically as each technique has its drawbacks, as shown above.

Here is an opportunity for you to evaluate some data on enzyme mechanisms for yourselves. Presented below are some observations made on the enzyme *chymotrypsin*. This enzyme has esterase activity, i.e. it hydrolyses a wide variety of esters (S24–, Unit 6).

$$R_1\!-\!\underset{\underset{O}{\|}}{C}\!-\!OR_2 + H_2O \rightleftharpoons R_1\underset{\underset{O}{\|}}{C}\!-\!OH + R_2OH$$

ester   *acid*   *alcohol*

It is envisaged that the enzyme, EH, is involved in an acyl intermediate (ECOR$_1$):

$$EH + R_1\underset{\underset{O}{\|}}{C}OR_2 \rightleftharpoons EHR_1\underset{\underset{O}{\|}}{C}OR_2 \rightleftharpoons ECOR_1 + R_2OH$$

enzyme

$$ECOR_1 + H_2O \rightleftharpoons EH + R_1COOH$$

This enzymic reaction operates best at around pH 7.

The formation of ECOR$_1$ would most probably require a strong nucleophilic group (S24–, Unit 6, and *Basic Organic Chemistry*, pp. 88–90), to be present as the part of the active site involved in the bond-breaking and bond-making steps.

Now examine the following pieces of evidence in turn. From the possible interpretations offered, pick out those that you consider to be most justified on the evidence presented up to that point; then proceed to the next set of evidence presented, and so on. Note that at certain points one may not necessarily be able to distinguish firmly between the alternatives given, as the process of evaluation should be progressive. In examining later evidence bear in mind evidence given earlier. When you have finished read the comments given on pp. 24–5. (If you are short of time on this Unit as a whole, you can treat this exercise in the way recommended in the Study Guide at the beginning of this Unit, p. 5.) Two additional points may prove useful:

1   In aqueous solution, the ionization of the terminal hydroxyl group of the amino acid serine (a primary alcoholic group) to give a seryl ion, a strong nucleophile, is virtually nil at pH 7.

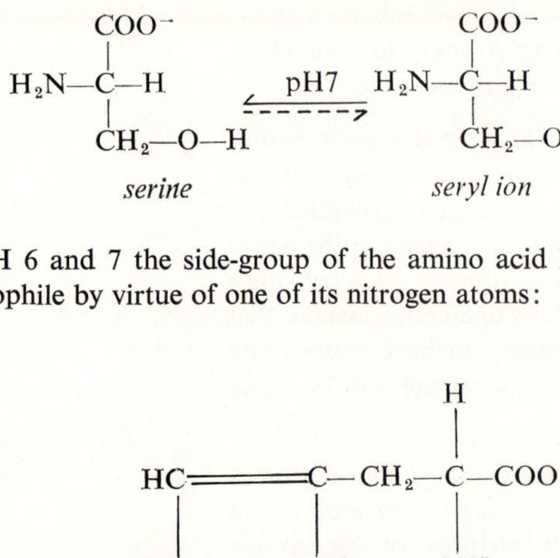

COO$^-$ COO$^-$

H$_2$N—C—H   pH7   H$_2$N—C—H

CH$_2$—O—H   CH$_2$—O$^-$

serine   *seryl ion*

2   Between pH 6 and 7 the side-group of the amino acid histidine could be a strong nucleophile by virtue of one of its nitrogen atoms:

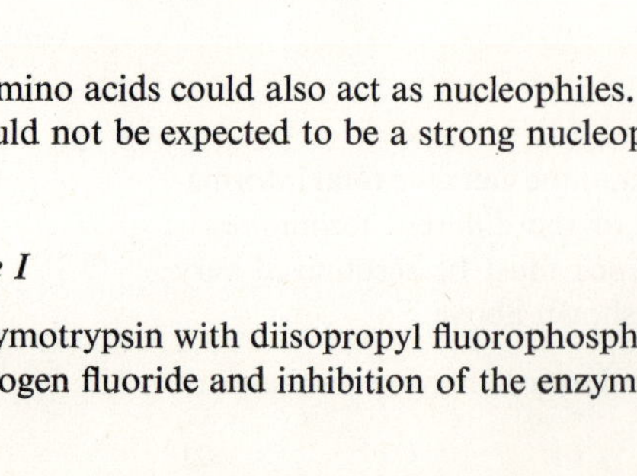

H
|
HC══C—CH$_2$—C—COOH
| |
N   NH   NH$_2$
_C_/
H
nucleophilic group

At pH 7 some other amino acids could also act as nucleophiles. But serine (i.e. —OH un-ionized) would not be expected to be a strong nucleophile at this pH.

Experimental Evidence I

(*a*)   Incubation of chymotrypsin with diisopropyl fluorophosphate (DFP) leads to production of hydrogen fluoride and inhibition of the enzyme.

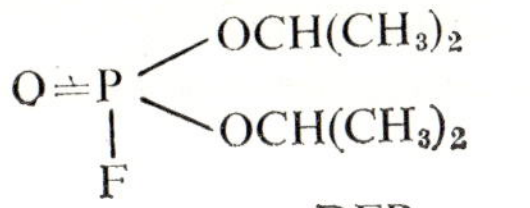

$$O \doteqdot P \underset{\underset{F}{|}}{\overset{\diagup OCH(CH_3)_2}{\diagdown OCH(CH_3)_2}}$$

DFP

(b)   The inhibition by DFP can be reversed by strong nucleophiles.

Interpretations of I

1   Hydrogen fluoride, a very corrosive chemical, destroys the enzyme.

2   DFP is a quasi-substrate (Section 2.4.2) which is binding to the active site that contains a nucleophile.

3   DFP is binding to the enzyme, not at the active site, affecting the active site as a secondary consequence of changing the shape of the enzyme.

Experimental Evidence II

(a)   When DFP in which the phosphorus atom is radioactive is incubated with chymotrypsin a radioactive protein is obtained.

(b)   1 mole of phosphorus (^{32}P) is bound per mole of enzyme, when inhibition is complete.

(c)   On hydrolysis of the inhibited enzyme with acid, o-diisopropyl phosphoryl serine can be isolated.

o-diisopropyl phosphoryl serine

$$H_2N-\underset{\underset{CH_2-O-P}{|}}{\overset{\overset{COOH}{|}}{C}-H} \quad \overset{O}{\underset{}{\|}} \underset{OCH(CH_3)_2}{\overset{OCH(CH_3)_2}{}}$$

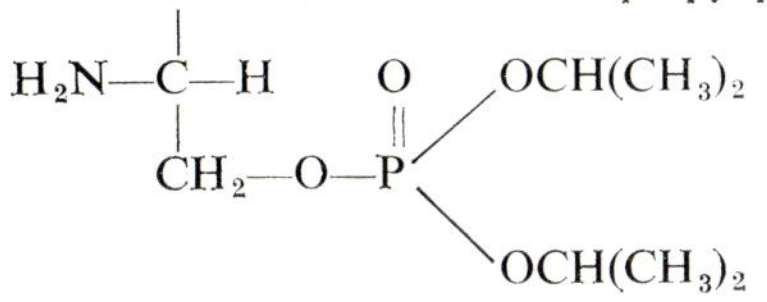

Interpretations of II

1   DFP is binding to the active site in which a serine side-group in the seryl form is a nucleophile:

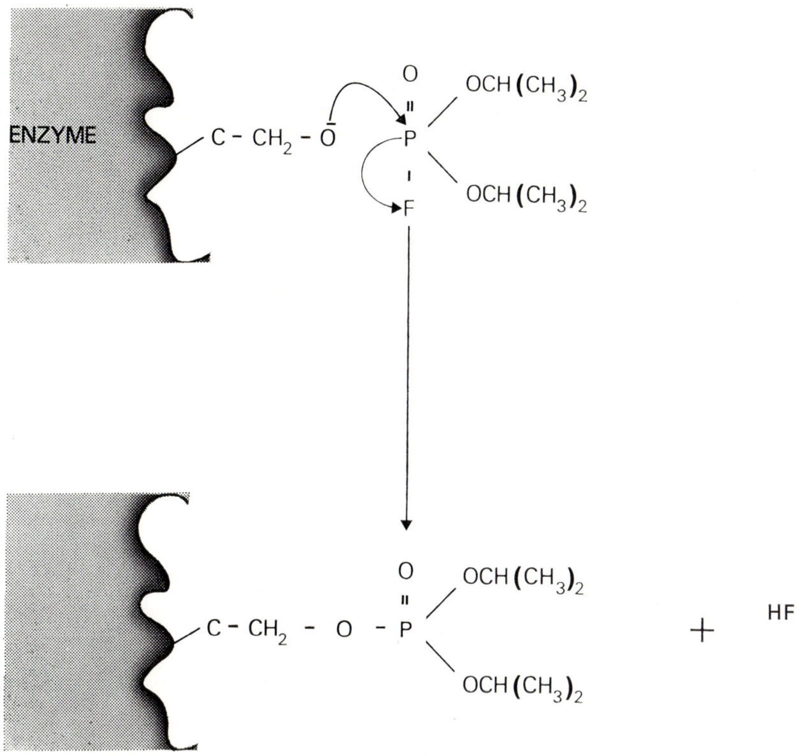

2   DFP destroys all serine side-groups in chymotrypsin (there are 29), thus causing a drastic change in the shape of the protein.

23

3 DFP binds to one particular serine side-group, not a side-group responsible for catalysis, causing a change in shape in the active site as a secondary consequence.

Experimental Evidence III

(*a*) The serine side-group attacked by DFP is shown to be only one of the 29 in chymotrypsin, that in position 195 in the chain (i.e. the 195th amino acid from the N-terminus).

(*b*) Reagents destroying the amino acid histidine inhibit the enzyme.

(*c*) pH studies reveal that a group with a pK_A of 6.6 is important in the catalysis.

(*d*) The 'free' amino acid (i.e. not in a protein) serine does not react with DFP.

Interpretations of III

1 One special serine side-group is a nucleophile at the active site.

2 A histidine side-group with a pK_A between 6 and 7 is the true nucleophile in the active site. The DFP reaction on the serine 195 side-group is an artifact; DFP is not a quasi-substrate.

3 The side-group of serine-195 is the true catalytic nucleophile, its activity depending on a histidine side-group.

Experimental Evidence IV

(*a*) A substrate of chymotrypsin, *p*-nitrophenyl acetate, can be shown to yield an acetyl enzyme as an intermediate.

(*b*) On hydrolysis of the acetyl enzyme, *o*-acetyl serine can be found.

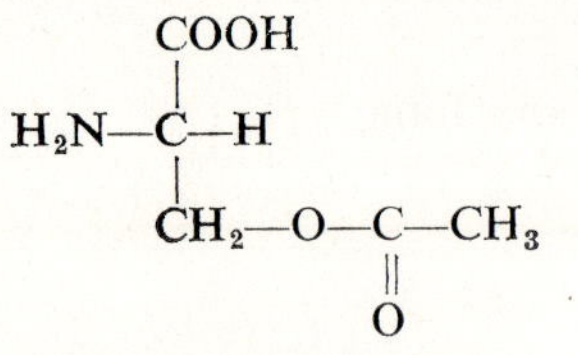

$$
\begin{array}{c}
\mathrm{COOH} \\
| \\
\mathrm{H_2N-C-H} \qquad\qquad \textit{o-acetyl serine} \\
| \\
\mathrm{CH_2-O-C-CH_3} \\
\quad\;\; \| \\
\quad\;\; \mathrm{O}
\end{array}
$$

(*c*) X-ray diffraction studies on crystalline chymotrypsin show that a histidine side-group is near serine-195 in the active site.

(*d*) Some other esterase enzymes react with DFP, and on hydrolysis *o*-diisopropyl phosphoryl serine can be found.

Interpretations of IV

1 The serine side-group is the true catalytic nucleophile.

2 The histidine side-group is the true catalytic nucleophile; the serine is an artifact.

3 The serine-195 and histidine side-groups interact in some way so as to bring about catalysis.

By now you should have formed some opinion as to the nature of the active site of chymotrypsin. Do you think a serine side-group is involved? If so, how? Now read the comments below.

Comments on I

Interpretation (1) is ruled out as the inhibition can be reversed (evidence I (*b*)). Either interpretation (2) or (3) could be correct; both go too far at this stage, since no change in shape of the enzyme has been noted nor is there any evidence that DFP is a quasi-substrate. Since, however, the reaction with DFP releases hydrogen fluoride, it is at least reasonable to suppose that wherever DFP is acting it has been attacked by a nucleophile:

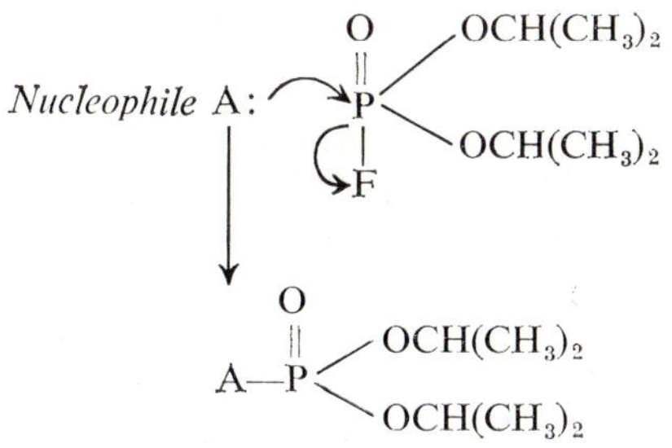

Comments on II

Interpretation (2) is virtually ruled out, as only 1 atom of phosphorus is bound per molecule of enzyme and yet there are 29 serine side-groups present (since 1 mole of phosphorus contains the same number of atoms as 1 mole of the enzyme does molecules; Avogadro's Number – S100, Unit 6).

Both interpretations (1) and (3) are possible at this stage. However, no experiments that could detect a change in the shape of the enzyme have been described, so any conclusions such as those in (3) are premature. From the facts known about the serine side-group, one would not expect it to be a strong enough nucleophile unless ionized. Ionization of this group is low at pH 7. Therefore hardly any of the interpretations seem fully acceptable at this stage. More data is needed.

Comments on III

Any one of the interpretations could be correct, but: Interpretation (1) is very vague – what is 'special' about this particular serine side-group?
If (2) is correct, then all data on DFP must be irrelevant to the mechanism of chymotrypsin.

Interpretation (3) fits the data so far, but it fails to explain how the serine side-group could be a nucleophile at the pH at which chymotrypsin operates (pH 7), particularly since free serine (evidence (d)) does not react with DFP.

Comments on IV

Interpretation (2) is unlikely to be correct, as the data on DFP and now from a substrate (IV (a) and (b)) both imply that the serine side-group is the catalytic nucleophile.

Interpretation (1), though probably correct, provides no mechanism for explaining the nucleophilicity of the serine side-group nor for the involvement of the histidine side-group.

Interpretation (3) seems the best, it providing a role for both histidine and serine side-groups, which are near each other (IV (c)).

Summary and conclusions

You should now appreciate some of the difficulties in evaluating data on enzyme mechanisms. From an original position where it seemed that the serine side-group could not be sufficiently nucleophilic it now appears that the cumulative evidence forces one to accept it as a nucleophile, at least for one such serine side-group in chymotrypsin and some other esterases. A histidine side-group is also involved – some suggestions as to how will be considered later (Section 2.5.4).

So you can see that to regard an enzyme as a lot of amino acids joined together, and therefore reflecting the properties of the individual amino acids when unjoined, is a misleading oversimplification. The three-dimensional nature of enzyme structure allows many interactions between the side-groups of the component amino acids, and it is these interactions that must form the basis of catalysis.

2.5 Catalytic Power

Study Comment

In this Section we examine some of the hypotheses advanced to explain enzymic catalysis. You are not expected to remember any of the detailed mechanisms but should be able to fulfil Objective 1 for those terms from this Section listed in Table A.

Even if the chemical nature of the active site of an enzyme is known, this in itself does not explain how catalytic factors of 10^9–10^{12} are achieved. This huge catalytic effect is termed the *catalytic power*.

There is no generally accepted explanation of catalytic power. Indeed, this subject is often the cause of many heated discussions at scientific meetings, through learned journals and in academic tea-rooms. We cannot present you with a series of well-substantiated facts nor with clearly defined hypotheses. However, no treatment of enzymes is complete without consideration of this key topic, so what we have done is briefly to review below some of the major hypotheses in this area.

2.5.1 Proximity and 'steering' effects

In a reaction involving two molecules (a bimolecular reaction), the molecules must collide in order to react. The rate of collision will depend on various parameters, including the concentrations of the reacting molecules (S100, Units 11 and 12). Substrates for enzymic reactions are generally in low concentrations, and it therefore seems reasonable to suppose that enzymes that catalyse reactions involving more than one substrate molecule increase the probability of collision between the substrate molecules by holding them near to each other in the active site. If, on this basis, we calculate the expected increase in collision and hence the rate of reaction for particular reactions this turns out to be small, in fact decreases are possible. This is probably because the enzyme itself is usually in such low concentration, on a molar basis, that the advantage gained by the proximity effect described above is offset by the low concentration of active sites. This proximity effect is therefore unlikely to contribute greatly towards catalytic power.

More recently (1970–1), some results from non-enzymic systems have been interpreted as indicating that the exact angle at which two molecules collide is critical to whether this collision leads to reaction or not. This had led to the idea, notably from Koshland and his co-workers, that enzymes may not merely bring molecules near to each other but may actually 'steer' the molecules into the precise position required for reaction, thus greatly catalysing the rate of reaction. This interesting idea caused a considerable stir among many chemists and biochemists alike. The controversy as to whether it is remotely true for enzymes or not is still current.

2.5.2 Acid-base catalysis

Many enzyme-catalysed reactions involve the movement of protons, that is acid-base equilibria. It is also known that pH affects the activity of enzymes. So it has been proposed that enzymes operate by 'general-acid–general-base catalysis' (that is, a non-protonated group is acting as a base and a protonated group as an acid). Studies on non-enzymic systems which model enzymic ones show that acid-base catalysis can occur, but the catalytic power in such model systems is way below that found in the enzymic systems.

An interesting variant of acid-base catalysis is that discovered by Swain and Brown. In 1952 they were examining a reaction that occurred in benzene solution, catalysed by a mixture of phenol and pyridine, but not by either alone. They concluded that the reaction was acid-base catalysed, pyridine being the base and phenol the acid. The reaction can be represented as follows:

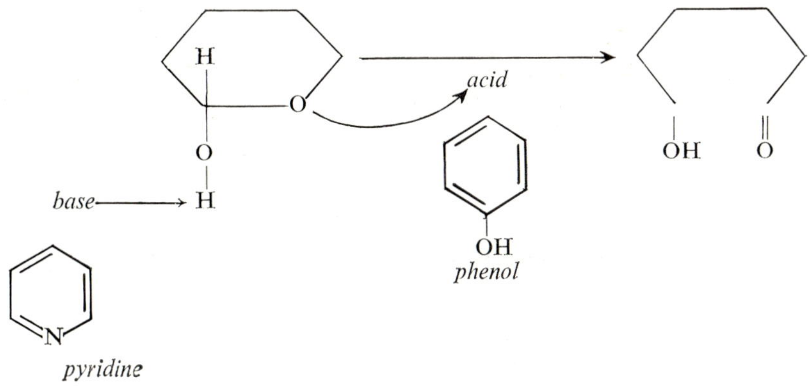

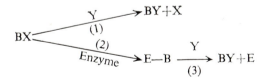

Presumably the acid- and base-catalysed steps would not occur together, each depending on a random collision with the molecule undergoing hydrolysis. Swain and Brown found that 2-hydroxypyridine, a much weaker base than pyridine and a weaker acid than phenol, catalysed the reaction 7 000 times as fast as pyridine plus phenol. Since 3-hydroxypyridine and 4-hydroxypyridine were nowhere near as efficient as 2-hydroxypyridine it appears that 2-hydroxypyridine is a powerful catalyst of this reaction by virtue of its structure enabling it to provide the acidic and basic groups synchronously:

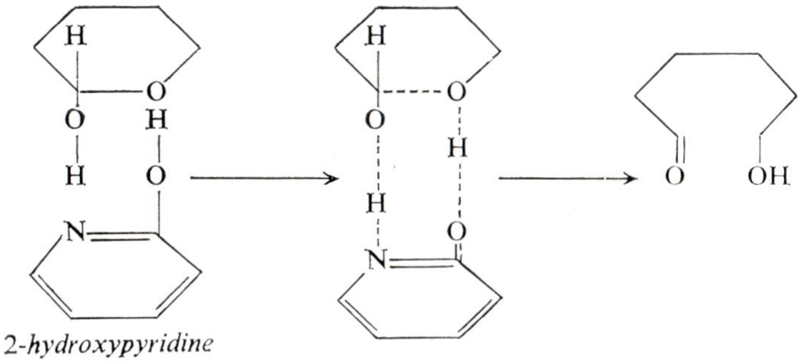

Though it is now hotly disputed whether such a synchronous event does occur in this system, it does provide an interesting model for how such 'push-pull' catalysis (the base 'pushing', the acid 'pulling') might occur in enzymes.

'push-pull'

2.5.3   Covalent intermediate catalysis

In some enzyme reactions, it has been shown that a covalent complex is formed between the enzyme and one of the reaction intermediates. Chymotrypsin, which you considered in Section 2.4.5, would be a case in point. This in itself does not prove that such complex formation increases the rate of reaction. For this to be true, in the scheme shown in Figure 4, both steps (2) and (3) must be faster than step (1).

Figure 4   Covalent intermediate catalysis; (1) shows overall uncatalysed reaction; (2) plus (3) show enzyme-catalysed reaction.

Compounds similar in structure to some thought to be involved in the formation of enzyme intermediates have been studied in order to examine possible catalytic effects of the formation of covalent intermediates. Though these model systems suggest that some catalysis is achieved, it is of a low order compared with enzymic

27

catalysis. It is therefore unlikely that the formation of covalent intermediates provides much of the catalytic power of those enzymic reactions that involve such intermediates.

2.5.4 Rack and strain; electronic strain

On the binding of a substrate to an enzyme the substrate may be distorted into an energetically unfavourable shape, thus introducing 'strain' into certain of the chemical bonds. Such strain may assist in catalysis, the bonds now being more susceptible to reaction (S24–, Unit 3). Some degree of strain catalysis may be occurring in lysozyme where, as you saw in *Cell Structure and Function*, pp. 254–8 (Figs. 10–19 and 10–20) the substrate is twisted into a strained 'half-chair' shape (S100, Unit 10).

'strain'

A variation of this idea is called the 'rack' hypothesis. In this hypothesis, it is considered that the enzyme itself is strained. Binding of the substrate leads to a change from the strained shape of the enzyme, and the strain energy is applied towards lowering the activation energy for reaction of the now 'tortured' substrate. These rack and strain hypotheses are shown diagrammatically in Figure 5.

'rack'

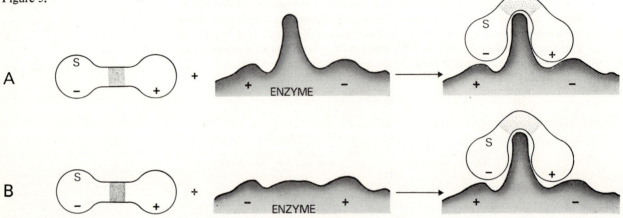

Figure 5 *Rack and strain.*

A) The reactive bond in the substrates (stippled area) is strained by the binding of groups on the substrate to complementary groups on the enzyme (+ and − indicate complementarity, not charge.)

(B) Rack mechanism. Binding of substrate leads to a change in the shape of the enzyme and hence strain in the substrate.

One of the problems of the rack and strain hypotheses is where the energy to produce the strained enzyme or substrates comes from. This is as yet no more easy to answer definitively than how catalysis occurs, and it is not possible for us to discuss this in detail at this juncture. Suffice it to say that several possible sources of energy exist, including energy derived from the binding of the substrate to the enzyme and energy stored in the enzyme originating from the specific folding of the protein into its three-dimensional shape.

Strain need not be geometric like that shown in Figure 5. Electronic strain, in enzyme or substrate, could also be important in catalysis. The abnormal reactivity of certain amino-acid side-groups may be explicable in such terms. It has been suggested that the serine side-group at the active site of chymotrypsin is a stronger nucleophile than expected (Section 2.4.5) because of electronic strain introduced into the terminal oxygen-hydrogen bond by the bonding of this hydrogen to a nearby histidine, this resulting in a larger partial negative charge on the oxygen:

electronic strain

$$-CH_2-O\overset{\delta^-}{}\ldots H\ldots N \qquad NH$$

serine hydroxyl   *histidine side-group*

However, at pH 7, the histidine side-group is probably not capable of causing appreciable dissociation of the serine hydrogen. It seems though that a nearby aspartic acid side-group renders the histidine side-group more basic, as shown in Figure 6.

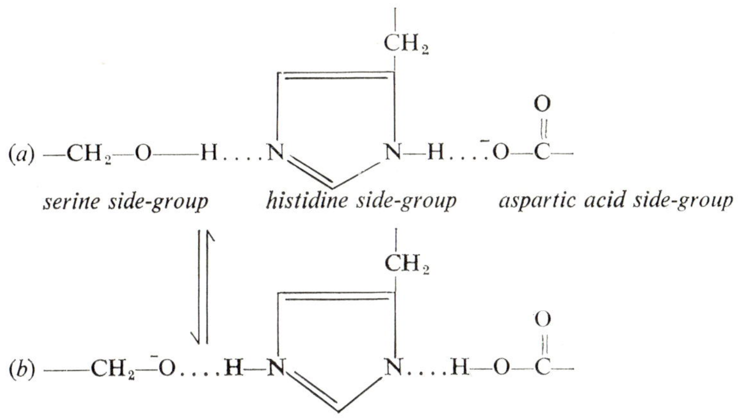

Figure 6 Active site of chymotrypsin. The bonding shown in (a) between the relevant portions of the serine, histidine and aspartic acid side-groups leads to the abstraction of a hydrogen from the serine side-group, resulting in (b) where the serine is now charged and hence a strong nucleophile.

Do not try to remember the details of this proposed active site structure. Just note how electronic strain due to interactions like those shown can lead to large perturbations from what are considered to be the 'normal' properties of amino-acid side-groups.

2.5.5   The micro-environment

We have now briefly reviewed some of the hypotheses currently advanced in an attempt to explain the catalytic power of enzymes. What we have said is undoubtedly a gross oversimplification of the situation, and it is improbable that any one of these hypotheses is absolutely correct, even for one enzyme. It is likely that several of the mechanisms operate together, each contributing towards catalysis, the relative contributions varying from enzyme to enzyme.

What in effect an enzyme does is provide a microscopic environment in which reactions can occur. In such an environment, redistribution of electronic charges can occur, substrates can be strained, and so on, all unimpeded by the surrounding medium. This last point is critical. In some instances, these enzymic reactions could be disrupted by water. In such cases, the access of water to the active site is limited by the precise tailoring of the three-dimensional structure of the protein. Such essentially non-aqueous environments are particularly important in certain reactions that take place in cellular membranes, and will be referred to in Unit 3 of this Course. By containing these microscopic 'alcoves' – the active sites of enzyme – one cell can encompass many different environments enabling it to execute a range of reactions at efficiencies as yet unmatched by the most modern laboratories.

Self-Assessment Questions for Sections 2.3 and 2.4 to 2.5.5

SAQ 4 *(Objective 3)*

The enzyme carboxypeptidase splits peptide bonds. The enzyme contains 1 atom of zinc. On removal of this zinc the enzyme is inactivated. Activity is restored by zinc or cobalt. X-ray analysis shows that addition of a substrate to the enzyme causes a tyrosine side-group in the protein to move by 12 Å to a position over the peptide bond in the substrate which is to be split. Acetylation of the enzyme inactivates it.

In the light of these data, consider whether the following statements are true, false, or possibly true but extrapolated beyond the data:

1   Zinc is involved in the structure of the protein, so that on its removal the protein is destroyed.

2   Koshland's induced-fit hypothesis is wrong.

3   A tyrosine-side group is involved in the catalysis.

4   Zinc is a cofactor.

5   Acetylation removes zinc.

SAQ 5 (*Objective 3*)

Consider the data given in *SAQ 4* plus the following: On acetylation of carboxypeptidase a modified tyrosine can be detected. X-ray data show that removal of zinc does not change the shape of the enzyme much. Now consider whether the following are true, false, or possibly true but going too far:

1   The role of tyrosine is to bind the zinc.

2   Zinc is involved in catalysis and not structure.

3   The tyrosine side-group is involved in catalysis.

SAQ 6 (*Objective 1*)

Are the following true or false?

1   The proximity effect explains enzyme catalysis.

2   A quasi-substrate is a compound that binds to the active site of an enzyme, as does the substrate, but the binding is more stable and does not lead to the formation of product.

3   The combination of an enzyme with its substrate is said to be a covalent intermediate.

4   The effect of pH on enzymic activity is completely explicable in terms of the pK_A values of the amino acids that comprise the active site.

5   The 'half-chair' shape of the substrate bound to lysozyme is evidence in favour of strain catalysis.

Now check your answers against those given on p. 39.

2.6   Isoenzymes

Study Comment

You do not need to remember the experimental details in this Section, but should remember the roles and functions of isoenzymes (Objective 4).

When we speak of a particular enzyme such as urease or lactic dehydrogenase, we are sometimes referring to a pure chemical entity, but more often than not we are describing something that has a particular catalytic activity. So urease is something that breaks down urea to ammonia and carbon dioxide, and lactic dehydrogenase catalyses the interconversion of lactic and pyruvic acids. Presumably all living organisms have many such reactions in common and therefore each possesses enzymes to catalyse these reactions. But are the enzymes that catalyse the same reactions in different organisms the same chemically? If the techniques developed for purifying proteins (Unit 1 of this Course) are applied to enzymes with the same catalytic role in different organisms, we generally find that they are chemically different. The differences increase as do other differences between the organisms—thus urease from man would be expected to be more like that from the chimpanzee than that from the haddock (S100, Unit 21).

However, when we speak of a particular enzyme from man, ape or fish, are we in each case speaking of a single unique type of enzyme? Is there sometimes more than one type of enzyme for a particular catalytic role? Do different organs in a complex organism have different enzymes for the same catalytic role? Why should they, it seems a waste?

Consider the data below. Treat it as you did that in Section 2.4.5. Choose the best interpretation at each stage, then go on. A commentary is given on pp. 32–3, which you should study after reaching your own conclusions based on the evidence and possible interpretations.

Evidence I

(*a*) See Figure 7.

(*b*) Five bands can be seen in Figure 7, labelled 1 to 5, in decreasing order of negativity (i.e. 1 is the most negatively charged and is therefore nearest the anode).

(*c*) Analogous patterns to that shown in Figure 7 can be obtained on examining the tissues of many other species of animals.

Interpretations of I

1 There are five different types of lactic dehydrogenase in humans, differently distributed between the different organs.

2 Several different proteins have been stained, not all of which are lactic dehydrogenase.

3 Only one type of lactic dehydrogenase is present but different impurities in the different tissues produce the different patterns.

4 Only two basic 'types' of lactic dehydrogenase are present, 1 and 5. The others are mixtures of 1 and 5.

Evidence II

The various types of lactic dehydrogenase (as seen in Fig. 7) were separated, purified and examined:

(*a*) Both type 1 and type 5 have molecular weights of 134 000.

(*b*) Treatment of type 1 or type 5 with a chemical that separates protein subunits gives proteins of molecular weight 34 000, with no enzymic activity.

(*c*) Figure 8 shows the results obtained on mixing pure type 1 and pure type 5 in sodium chloride, then freezing and thawing. (If no freezing–thawing step is done, then just two bands corresponding to types 1 and 5 are observed.)

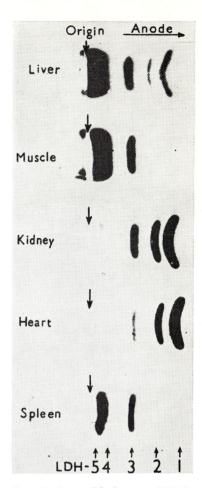

Figure 7 *Lactic dehydrogenase (LDH) in human organs. Extracts from the organs indicated were run by starch gel electrophoresis (Unit 1 and TV 3) then stained for lactic dehydrogenase. This was done by incubating the gel after electrophoresis with the substrates of the enzyme plus a dye, which on reaction with a product of the enzymic reaction gives an insoluble coloured product. Since this product is insoluble it is 'fixed' on the gel where it is produced, thus pinpointing the enzyme. The origin, that is the point on the gel where the organ extracts were applied, is indicated by the vertical arrow. The anode is on the right, the cathode on the left (S100, Unit 9).*

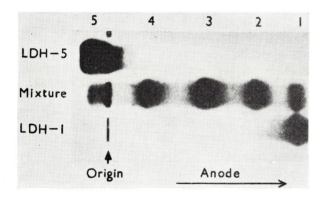

Figure 8 *Starch gel electrophoresis of type 1, type 5 and sodium chloride-frozen-thawed mixture. Stained for lactic dehydrogenase (LDH).*

Interpretations of II

1 Since types 1 and 5 have the same molecular weight they are the same enzyme. Low molecular weight impurities give them different electrophoretic properties.

2 Types 1 and 5 are different enzymes. 2, 3 and 4 are mixtures of 1 and 5 in differing proportions.

3 Type 1 is composed of four identical subunits – BBBB. 5 is composed of four identical (but different from those of 1) subunits – AAAA. Types 2, 3, 4 are mixtures of A and B in differing proportions.

Evidence III

(*a*) Antibodies produced against pure type 5 inhibit the enzymic activities of types 5, 4, 3, 2 and 1 by 86%, 68%, 41%, 23% and 0%, respectively. (Antibodies

31

are proteins which are produced by injecting a protein from an animal of one species into an animal of another species. Proteins, called antibodies, are then produced by the recipient animal, which combine specifically with the 'foreign' protein which was injected.)

(*b*) Genetic studies on humans suggest that two genes exist for lactic dehydrogenase.

(*c*) All five human types of lactic dehydrogenase have the same molecular weight.

Interpretations of III

1 Only two different enzymes are present – 1 and 5. Types 2, 3, 4 are mixtures.

2 Type 1 consists of 4 subunits – BBBB (or B_4); type 5 consists of 4 subunits – AAAA (or A_4). Type 2 is B_3A, Type 3 is B_2A_2, Type 4 is BA_3 (order irrelevant; e.g., BA_3 and A_3B are the same).

Evidence IV

(*a*) Type 1 is inhibited by low levels of pyruvic acid, type 5 is not.

(*b*) Type 1 is inhibited by lactic acid, type 5 is not.

At this stage review all the evidence and interpretations. Bear in mind that the physiological role of lactic dehydrogenase is to reduce pyruvic acid to lactic acid:

$$\underset{\text{pyruvic acid}}{\overset{\displaystyle CH_3}{\underset{\displaystyle COOH}{\overset{|}{\underset{|}{C=O}}}}} + NADH_2 \rightleftharpoons \underset{\text{lactic acid}}{H-\overset{\displaystyle CH_3}{\underset{\displaystyle COOH}{\overset{|}{\underset{|}{C}}-OH}}} + NAD$$

Also consider what you know about the occurrence of lactic acid in muscle (S100, Units 15 and 16). Refer again to Figure 7 and, from the patterns found for muscle and heart, try to determine the consequences of these different patterns to the metabolism of pyruvic acid in the two types of organ. Do not spend more than about 20 minutes trying to do this last part of the exercise. Then read the comments below.

Comments on I

Interpretation (1) is best, as it merely reflects what can be seen from Figure 7.

(2) Is unlikely as the staining technique depends on the reaction catalysed by lactic dehydrogenase.

(3) Is unlikely as the pattern is similar among many organisms. This is unlikely to be due to 'impurities' being the same in different organisms.

(4) Is possible but assumes too much. 'Mixtures' is vague.

Comments on II

Interpretation (3) is best. The idea of 4 subunits is compatible with the dissociation from a molecular weight of 134 000 to one of 34 000. One is not justified, however, in concluding that the 4 subunits are identical because of the similarity in molecular weight. The identity is supported by (*c*), as mixing of A and B to give tetrameric structures (i.e. containing 4 subunits) could give 3 'new' types— A_3B_1, A_2B_2, AB_3.

(2) Is also possible at this stage. If 2, 3 and 4 are mixtures of 1 and 5 they would have proportionally higher molecular weights than 1 and 5.

(3) Is possible but poorly reasoned, as similarity in molecular weight is an insufficient criterion of identity.

Comments on III

Interpretation (2) is almost certainly correct. Evidence (*c*) rules out (1) and incidentally II (2) as the molecular weights of the mixtures should be greater than either types 1 or 5. Two genes, each producing one type of subunit (A or B) could give rise to five possible tetramers. The suggested composition of types 1 to 5 is compatible with the inhibition by antibodies against type 5. These anti-bodies presumably are specific against subunit A and therefore inhibit the different types of the enzyme in proportion to the number of these subunits (A) that each type of enzyme contains.

Comments on IV

You probably concluded correctly that the structures of types 1 to 5 are those suggested in III (2).

What about the metabolism of pyruvic acid in heart and muscle? Since type 1 is inhibited by pyruvic and lactic acids, this must be a property of subunit B. So the types of enzyme will be susceptible to this inhibition in proportion to their content of subunit B. Inhibition by pyruvic and lactic acids would lead to an accumulation of pyruvic acid in tissues having predominantly lactic dehydro-genase types containing B.

This pyruvic acid will then feed into the Krebs cycle and be oxidized to carbon dioxide and water. Thus tissues containing lactic dehydrogenase with mainly B subunits (types 1 and 2) will be adapted to aerobic metabolism. Those with a preponderance of A – containing lactic dehydrogenase (types 4 and 5) – will tend to accumulate lactic acid and be anaerobic. From Figure 7 you can see that, on this basis, heart must be aerobic while muscle could be anaerobic – this is indeed so. This is not to say that this is the sole reason why these tissues are aerobic or anaerobic, just that their lactic dehydrogenase is adapted to these forms of metabolism.

This close relationship between structure and function is elegantly borne out by studies on the lactic dehydrogenase types of the breast muscle of different birds. Birds which carry out sustained flight, such as the stormy petrel and swift, are found to have mainly B subunits in this tissue, thus enabling long flights without a build-up of lactic acid which would lead to cramp (S100, Units 15 and 16). Birds which need only fly in short bursts, such as the rather sedentary domestic chicken, are found to have mainly A subunits.

Hitherto, we have called the five different forms of lactic dehydrogenase, 'types'. The name usually applied is 'isoenzymes', or 'isozymes'. Many other enzymes from various sources can be shown to consist of several isoenzymes. The number of component isoenzymes varies from enzyme to enzyme and from source to source. Where a charge difference exists between the isoenzymes, then electro-phoresis can readily demonstrate their existence, though other separative techniques could be used if no charge difference occurs. Presumably, as with lactic dehydrogenase, each isoenzyme of an enzyme fulfils some specific physio-logical role. We return to these roles when discussing the control of cell meta-bolism in Unit 5 of this Course.

Within the same species of organism, variations in isoenzyme patterns between individuals can occasionally be found. Such variation, for example in fruit-flies and some plants, is currently under active study throughout the world. Not only will such studies yield information concerning the physiological roles of the various isoenzyme systems but should give valuable data on the evolution of populations of organisms (S100, Unit 19).

2.7 Social Applications of Knowledge about Enzymes

Study Comment

You should as a result of reading this Section be able in part to evaluate the relevance and consequences of studies on enzymes (Objective 5). You may find that being able to remember some examples helps you in fulfilling this objective.

By now it should be apparent that a considerable amount is already known about enzymes:

1 Enzymes are catalysts found in all living cells.

2   Enzymes are composed mainly of protein, but also can contain non-protein cofactors.

3   Cofactors and substrates fit into an accurately shaped active site, hence enzyme specificity.

4   Non-substrate compounds which bind to active sites can inhibit enzymes.

5   Different tissues within the same organism can contain variants of enzymes with the same catalytic activity—isoenzymes.

As yet, the application of such knowledge to everyday life is limited to the field of medicine, and a few industrial applications. We shall now consider some of these present-day applications and the directions in which such knowledge may be applied in the future.

2.7.1   Clinical enzymology in diagnosis

There are basically two phases in curing disease—diagnosis and therapy. In the field of diagnosis, enzyme studies are firmly established.

Since individuals only vary slightly from each other in what we might call their normal levels of enzymes, by measuring the levels of various enzymes in many tissues of many individuals we can build-up a catalogue of normal ranges for these enzymes and tissues. Then, by comparing these with individuals known to be suffering from well-defined diseases, we can observe how different diseases do or do not affect these enzymes. This data can then be used to assist in diagnosing these diseases in other people.

In principle, many tissues can be examined in this way but by far the easiest to obtain from a living person is blood. Following removal by centrifugation of red blood cells from blood, the remaining fraction, the serum, can be assayed for enzymes. As sensitive enzyme assays have been developed only small blood samples are needed, and the patient can provide these without fearing, as comedian Tony Hancock once did, that they are losing 'nearly an armful'.

Many diseases do lead to changes in the levels of enzymes in the serum. Though the exact reasons for these changes are not always known, they are nevertheless useful in diagnosing diseases, or at least in helping confirm a diagnosis. Often taking successive measurements over a period of time can be used to follow the course of a disease; the level of serum alanine aminotransferase (an enzyme that transfers the amino group of alanine to oxoacids) can be used to monitor the course of infectious hepatitis (a liver disease). Since, however, several different diseases can often lead to essentially the same change in the level of a particular enzyme, it is more valuable to have data on several enzymes than on one alone. For example, diseases of the liver, heart and kidneys can all lead to increases in the serum level of lactic dehydrogenase. In the case of lactic dehydrogenase (and some other enzymes), by studying the isoenzyme pattern, rather than the level of enzyme activity in the serum, we can obtain more definitive information as to the cause of the increases. From Figure 9 you can judge for yourselves how a heart attack can effect changes in the serum lactic dehydrogenase. Note the increase in isoenzyme 1. Compare Figure 9 with Figure 7. You can now see that the heart attack has probably led to some heart protein being released into the blood. So the serum pattern (Fig. 9) indicates some damage to the heart.

Since the main object of accurate diagnosis is to detect the disease as early as possible so as to heighten the chances of successfully curing the patient, such techniques are very useful.

The pattern of lactic dehydrogenase isoenzymes in serum can also give useful information on patients who have received a heart or liver transplant. Where the transplanted organ is about to reject, this is heralded by increases in the serum level of the isoenzymes characteristic of the organ in question. Similarly, early warning of cancer of the intestine or uterus is given by an increase in the serum level of lactic dehydrogenase isoenzymes characteristic of those tissues, before the tissues themselves show any signs of malignancy.

In certain 'metabolic diseases', diagnosis by enzyme assay is invaluable. 'Metabolic diseases' often arise from some inherited defect such as the lack of a particular enzyme, as a result of a specific mutation (S100, Unit 19). Babies suffering from 'galactosaemia' cannot metabolize galactose properly. If untreated this

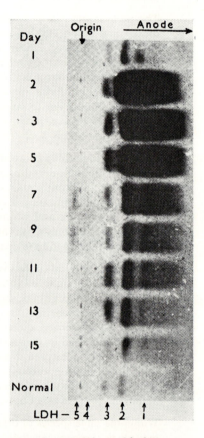

Figure 9   Serum lactic dehydrogenase isoenzymes following a heart attack on day 1. Starch gel electrophoresis pattern.

metabolic diseases

34

has serious consequences: liver damage and mental deficiency. The cure is relatively simple if the disease is spotted early enough – feeding the baby a galactose-free diet. As the disease arises from the lack of one of the enzymes involved in galactose metabolism, it is readily diagnosed.

It is not only during disease that changes in enzyme levels occur. Large changes occur in serum enzyme levels during pregnancy. This is not particularly useful as a diagnosis of pregnancy, but, where the pregnancy is abnormal, failure of the enzyme changes to occur can give early warning.

2.7.2   Chemotherapy

Having diagnosed a disease, the next step is to try to cure it. When the disease is caused by a 'foreign' organism such as a bacterium or a virus, or is due to an abnormal type of cell as in cancer, the main line of treatment is to kill or curtail the growth of the 'invader' without harming the patient. Frequently drugs are employed and such treatment is called *chemotherapy*.

Knowledge about enzymes is potentially of great value here. For example, if an invading bacterium was known to contain an enzyme which was unlike any in the host (patient), then ideally we could design a chemical which specifically inhibits this unique enzyme and hence attacks the bacterium alone. Such an approach involves two main principles:

1   The enzyme must be unique to the 'invader'.

2   The chemical must affect this unique enzyme and not affect any of the host's metabolism, or else undesirable side-effects could ensue.

In practice, these two requirements are hard to meet. Since most biochemical reactions are common to virtually all organisms, similar enzymes will be present in host and invader. In addition, it is very difficult to design chemicals which only inhibit one specific enzyme. But minor chemical, if not catalytic, differences do seem to exist between similar enzymes from different sources and so, given time, one should be able to attack specifically any organism. Fortunately some enzymes 'unique' to bacteria do seem to exist and it is probable that some antibiotics operate against these. You have already come across one such example – sulphanilamide. (The mode of action of this antibiotic is explained in *The Chemistry of Life*, p. 97, and in *Cell Structure and Function*, p. 243.)

The knowledge of exactly how sulphanilamide acts came after its use as a therapeutic agent. The same is true for nearly all drugs. Undoubtedly the more one learns about enzymes the more it should be possible to 'tailor' specific inhibitors and kill invading organisms. This seems more logical than a 'trial and error' testing approach. It may be, however, that trial and error is more efficient both in terms of time and money than a more rational progress from basic research to drug design. A combination of both approaches is currently employed.

It is also important to be aware of the dangers of certain possible applications of research into enzyme inhibitors. DFP, mentioned in Section 2.4.5, is a *nerve gas*, that is, a substance which inhibits the normal functioning of nerves. It, and related substances, have this property by virtue of their ability to combine with the serine at the active site of acetylcholinesterase (an enzyme vital to nerve function). These nerve gases were known as poisons before their mode of action was understood. It would be unwise to assume that detailed knowledge on enzyme inhibition would not be used in the future for harmful purposes. To give but one example: there is currently an interest in producing drugs to control certain mental illnesses. Such substances, if misused, could probably be very dangerous, and some scientists fear that governments would be interested in such drugs to 'pacify' enemy troops, etc. In such a case, as in others, the responsibility of scientists with respect to the politics of how their researches are applied or not is an issue which must concern everyone.

nerve gas

Though 'tailoring' of specific enzyme inhibitors for clinical purposes is as yet in its infancy, enzymes themselves have sometimes been used clinically. An enzyme capable of digesting the connective tissue between cells can be injected along with drugs so as to increase the penetration of the drugs. Recently the use of an enzyme to treat some forms of cancer has gained much prominence.

Certain cancer cells, including some leukaemia cells, cannot synthesize their own supply of the amino acid asparagine. They therefore depend on a supply from the normal cells of the body. These cancer cells are therefore sensitive to an enzyme that destroys asparagine. Such an enzyme, asparaginase, which can be isolated from bacteria grown in bulk, has shown some beneficial effects against certain types of cancer in clinical trials. This promising agent is currently being further investigated.

2.7.3 Industrial and household uses

As enzymes as a whole can catalyse a wide variety of chemical reactions with high specificity and efficiency, it would seem that purified enzymes might be of great use in the chemical industry. They have been used in some processes, such as the manufacture of antibiotics, but, as yet, their use has not been widely exploited. There are probably many reasons for this, including the difficulty and high cost of obtaining large quantities of pure enzymes and their inherent instability (TV 6).

In the food industry, however, enzymes are widely used to alter the consistency, taste and appearance of food. Enzymes are employed in the baking of bread, the production of cheese, the clarification of wines, the storing of beer, the making of liquid fruit centres such as those in sweets or chocolates, and in many more processes. The enzymes used are frequently very impure, as it is often enough to add a crude extract of the source of the enzyme to ensure sufficient catalysis. It is significant that enzymes are now used in the baking, brewing and dairy industries, since whole organisms, which of course contain enzymes, such as yeasts and fermentative bacteria, have been traditionally used in these industries for thousands of years. The use of pure enzymes, or partially purified ones, has not superseded the use of yeasts and bacteria, but addition of enzymes allows further varieties of food to be prepared under controlled conditions. Further developments in this field are inevitable and the use of enzymes to enhance or alter the flavour of foods is of interest to the scientist and gourmet alike – not that the two characters are mutually exclusive!

Though most of these processes are done at the industrial level, some enzyme preparations can now be used domestically. One traditional way of tenderizing beef is to let it hang in the cold for a few weeks. This tenderizing occurs owing to the action of proteolytic enzymes (enzymes that split peptide bonds in proteins; Unit 1 of this Course) present in the meat. It is possible to tenderize meat much more rapidly by addition of proteolytic enzymes, such as papain, and powders containing *papain* can be bought for this purpose. The modern cook using papain is in essence doing what South American Indians have been doing for hundreds or thousands of years, as they tenderize meat by wrapping it in the leaves of the *papaya* plant, the plant from which papain is obtained and from which it derives its name.

There is another more recent encroachment of enzymology into the household and indeed into popular folk-lore. Enzymes are now widely heralded as those things that get rid of 'biological stains'. This use of enzymes in washing powders, in fact, is quite logical. After all, if such enzymes help you digest an egg in your gut, why shouldn't they go to work on the egg on your tie? The principle is fine, as yet there seems to be some doubt as to whether the inclusion of enzymes in washing powders improves them or not. There are also some suggestions that some people may be adversely affected by large quantities of such powders (probably as a result of allergic reactions). This is a problem in some factories where the powders are manufactured.

Undoubtedly, as man's knowledge of enzymes increases over the next few decades, this will lead to more clinical and industrial applications. Both in terms of research and everyday life, what one of the founders of modern biochemistry said, forty years ago, is still true today: 'It is, I think, difficult to exaggerate the importance to biology, and I venture to say to chemistry no less, of extended studies of enzymes and their action.' – Sir Frederick Gowland Hopkins (1932).

Self-assessment Questions for Sections 2.6 to 2.7.3

SAQ 7 (*Objective 4*)

Are the following true or false?

Isoenzymes are of interest because they:

1 Help to prove the theory of evolution.

2 Assist in clinical diagnosis.

3 Are currently useful in chemotherapy.

4 Illustrate structure-function interrelationship.

5 Help birds fly.

6 Complicate enzyme purification.

SAQ 8 (*Objective 5*)

How much has knowledge about enzymes assisted (or is likely to assist) the following in the near future (Nil, Slightly, Considerably, Absolutely):

(*a*) A cure for cancer.
(*b*) Diagnosis of disease.
(*c*) Cheaper washing powders.
(*d*) Manufacture of antibiotics.
(*e*) Design of nerve gases.
(*f*) Decreasing pollution.
(*g*) Improving food.
(*h*) Synthesizing new chemical elements.
(*j*) 'Spare-part' surgery.

Appendix 1 (White)

pK_A
An acid may be defined as a substance that dissociates to give a proton and a *conjugate base*, that is, in the dissociation of a weak acid (S100, Unit 9) HA:

$$HA \rightleftharpoons H^+ + A^-$$

A^- is the conjugate base.

The equilibrium constant for the dissociation,

$$K_A = \frac{[A^-]\,[H^+]}{[HA]}$$

Therefore, $$\frac{1}{[H^+]} = \frac{1}{K_A} \cdot \frac{[A^-]}{[HA]}$$

Taking logarithms, $$pH = pK_A + \log_{10}\frac{[A^-]}{[HA]}$$

where pK_A is $-\log_{10}K_A$.
Where HA is 50 per cent ionized, $[A^-]=[HA]$ and thus

$$pH = pK_A + \log_{10}1 = pK_A$$

Therefore the pK_A for an acidic group is the pH at which that group is 50% ionized.

You should remember this definition of pK_A but you do not need to remember the above derivation.

So pK_A gives one a useful means of comparing the relative strengths of acidic groups; the stronger the acid the lower its pK_A.

So trichloracetic acid ($pK_A = 0.65$) is a stronger acid than acetic acid ($pK_A = 4.75$) which is a stronger acid than the ammonium ion ($NH_4^+ \rightleftharpoons NH_3 + H^+$) ($pK_A = 9.21$), since the lower the pK_A the lower the pH at which the acid is ionized and hence a proton donor.

Many compounds have more than one ionizable group and hence more than one pK_A. Many compounds of biological importance have two or more ionizable groups. For example, all amino acids have at least two ionizable groups, the carboxyl and amino groups:

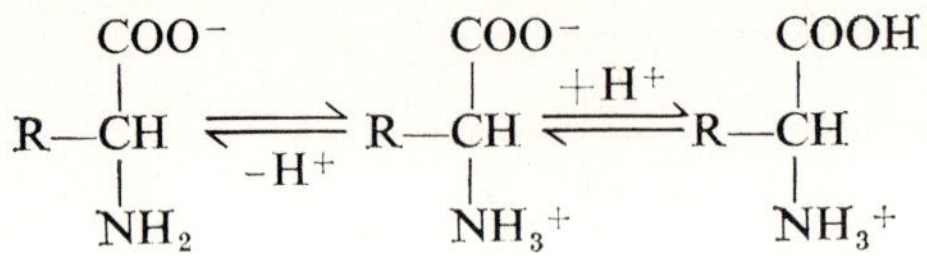

Some amino acids also have ionizable side-groups (*Cell Structure and Function*, p. 188), and these side-groups contribute to the three-dimensional structure, and hence function, of proteins (Unit 1 of this Course, and this Unit).

Pre-Unit Assessment Test: Answers and Comments

1 False. A catalyst increases the rate of a chemical reaction by participating in the reaction. However, at the end of the reaction the catalyst is regenerated.

2 True.

3 False. Though different enzymes exist, which taken together can catalyse a wide variety of reactions, any one enzyme can only catalyse one type of reaction among a narrow range of substances. This is termed *enzyme specificity*.

4 False. All enzymes so far isolated are, at least mainly, composed of protein. This had led to enzymes being regarded as protein catalysts.

5 True.

6 False. They can only cause an increase in the rates of reactions which are thermodynamically possible. They cannot alter the equilibrium position of a reaction, only increase the rate at which it is reached.

7 True.

8 False. Most enzymes can only catalyse reactions within a narrow range of pH, the so-called optimum pH. This pH is different for different enzymes.

9 (c) is the best alternative, since *some* substances chemically related to substrates can often bind to the active centre and inhibit reactions.

(a) is untrue as if A′ were a coenzyme it would if anything assist the reaction.
(b) is untrue as if A′ is not a substrate, A still is and the reaction should continue.
(d) is untrue as if A′ were a substrate both it and A would be acted upon and the reaction on A would not cease, although it could slow down due to competition with A′.

10 (a) is the best alternative as, although B is chemically similar to A, enzymes, being very *specific*, can often distinguish between closely related compounds.

If you were wrong on more than three of these ten questions you are advised to read *The Chemistry of Life*, pp. 88–100, before proceeding with this Unit.

Self-assessment Answers and Comments

SAQ 1

(*a*) (vi) (4). B absorbs ultra-violet light.

(*b*) (iii) (3); cannot use ultra-violet absorption as both B and G absorb.

(*c*) (iv) (3) or (iii) (3). N is yellow; L reacts with dye to give blue colour.

(*d*) (x) (2) or (x) (7). Cannot use ultra-violet absorption as both B and G absorb.

(*e*) (x) (7) or (vi) (4) or (iii) (3). B absorbs ultra-violet light; L reacts with dye to give blue colour.

(*f*) (viii) (1) or (vi) (4). C is a gas. G absorbs ultra-violet light.

(*g*) (vi) (4) G absorbs ultra-violet light. Both L and Z react with dye—therefore this cannot be used.

(*h*) Since C and Q are both gases and S and P are both solids giving non-absorbing solutions, we could only use a coupled assay, that is, take samples from (*h*) and measure C or Q or S or P present at various times. There are many possible ways of doing this: for example, any reaction involving S(e) or reaction (a) in reverse would measure P. Therefore (i) is the answer.

SAQ 2

(*b*) is correct.

(*a*) is incorrect. Observations made on an enzyme cannot be 'due to' an equation. The equation is merely consistent with the behaviour of the enzyme.

(*c*) is a possibility for some enzymes but unlikely to be true for all. This also does not explain the initial increase in rate.

(*d*) is untrue. Since the rate is increasing with increasing substrate it cannot be due to the substrate being depleted.

SAQ 3

1   False.

2   False. What is a vitamin for one organism is not necessarily a vitamin for another.

3   True.

4   False. They are often converted to cofactors.

5   True.

SAQ 4

1   False. Can be reversed by zinc or cobalt.

2   False. If anything, this supports the hypothesis (Section 2.3).

3   Possibly true, but going too far.

4   True. As it is needed for activity it is a cofactor by definition (Section 2.4.1).

5   Possibly true, but going too far. It could inhibit for a variety of reasons.

SAQ 5

1   Possibly true, but going too far. Seems unlikely in view of other possible roles of the tyrosine side-group (see below).

2   True. No large change in shape occurs on removal of zinc, but activity is lost. Therefore zinc is more likely to be involved in catalysis than structure.

3   Possibly true, but going too far. Movement of the tyrosine side-group over to the bond to be broken is not proof that it helps break the bond. Similarly inhibition by acetylation of the tyrosine side-group is also not conclusive proof of its catalytic role.

SAQ 6

1   False (Section 2.5.1).

2   True (Sections 2.4.2 and 2.4.5).

3   False (Section 2.5.3).

4   False (Section 2.4.3).

5   True (Sections 2.4.4 and 2.5.4).

SAQ 7

1   True, but only in a trivial sense. The existence of isoenzymes argues for evolutionary pressures (S100, Units 19 and 21).

2   True (Section 2.7.1).

3   False—only useful in diagnosis (Section 2.7.1).

4   True.

5   Oversimplification (Section 2.6).

6   True. Ask yourself: is a pure enzyme one protein or several isoenzymes? Or, is one pure enzyme just one isoenzyme of several?

SAQ 8

The answers are a matter of opinion, but:

(*a*) Slightly or Considerably.

(*b*) Considerably.

(*c*) Nil – cheaper?

(*d*) Slightly or Considerably.

(*e*) Considerably – where *designed*.

(*f*) Nil or Slightly.

(*g*) Nil or Slightly – 'improvement' is debatable.

(*h*) Nil.

(*j*) Slightly.

Acknowledgements

Grateful acknowledgement is made to the following for illustrations used in this Unit:

Figures 7 and 9: Academic Press for A. L. Latner, *Isoenzymes in Biology and Medicine;* Figure 8: American Association for the Advancement of Science for C. L. Markert, 'Lactate dehydrogenase isozymes: Dissociation and recombination of subunits' in *Science;* **140,** 1963.

Biochemistry

1 Biological Macromolecules
2 Enzymes
3 Cell Energetics
4 Metabolic Pathways
5
6 Regulation of Cell Processes